PROBLEM SOLVING

with | Ratio and Proportion

Doreen Nation
Sheila Siderman

Creative Publications®
A Tribune Education Company

Acknowledgments

Editorial Services Siderman Nation Publishing Services Inc.

Project Editor Janet Pittock

Design Director Karen Stack

Design O'Connor Design

Cover Illustration Lisa Manning

Illustration Jim Dandy

Production A. W. Kingston Publishing Services, LLC

Manufacturing Dallas Richards

ISBN 0-7622-1255-1

Catalog #10874

Customer Service 800-624-0822 or 708-385-0110

http://www.creativepublications.com

1 2 3 4 5 6 7 8   ML   06 05 04 03 02 01 00

Contents

Overview

What are the Goals of **Problem Solving with Ratio and Proportion**?

Throughout their lives, students have a need for proportional reasoning, in school and out of school, with numerical and geometric applications. **Problem Solving with Ratio and Proportion** provides students with many situations that represent both types of problems.

Students using **Problem Solving with Ratio and Proportion** will learn

- to recognize which problems can be solved by proportional reasoning and how to apply that reasoning.

- to compare data using rates and unit rates.

- to use comparisons with enlargements, reductions, scale drawings, maps, and blueprints.

- to recognize that the idea of similarity is vital to a comparison of physical representations, whether enlargements, reductions, scale drawings, maps, or blueprints.

- that reasoning applied to physical models can be applied to numerical relationships.

- that problems involving ratio and proportion can often be solved using problem-solving strategies students already know.

One of the goals of all mathematics education is for students to become successful problem solvers. Toward this end, **Problem Solving with Ratio and Proportion** places equal emphasis on content and process. Problems are designed so that in many cases, students will be able to choose from a variety of approaches. Often, there is no single best approach. We recommend that as problems are solved, all student suggestions for solution methods be considered. This will reinforce students' understanding that various approaches can often be used to solve the same problem. It will also encourage students to feel comfortable making suggestions. Students are more likely to participate if they understand that incorrect solution paths will be valued for what can be learned from them.

This program is intended for use by middle school students who have achieved a basic understanding of the ideas of ratio as a comparison. It is assumed that students have studied ratio and proportion topics, such as enlargements and reductions, scale drawings, maps, and blueprints. Therefore, the program does not teach ratio and proportion topics, but rather provides problem situations in which students can apply their understandings to the concepts.

How is **Problem Solving with Ratio and Proportion** Organized?

Sections

Each of six types of proportional reasoning situations is presented in its own section: *Rates*, *Unit Rates*, *Enlargements and Reductions*, *Scale Drawings*, *Maps*, and *Blueprints*.

Each section includes integrated information about instruction, applications, and assessment. Within each section there are three levels of application activities. In order of increasing difficulty, they are: *Try It Out*, *Stretch Your Thinking*, and *Challenge Your Mind*.

Instruction

Each section begins with a two-page teacher-oriented *Introduction*. There is a discussion of the types of problems in the section and the mathematics that students will encounter. This is followed by notes on various problem-solving strategies and solution paths that can be used for the problems in the section. Each introductory section also contains five *Thinking About…* problems that focus students' attention on number sense and estimation as they apply to the problems they will encounter.

The first problem in each section, *Solving Problems with…*, will focus students' attention on real-world problems that can be solved using that particular skill. Throughout the program, a variety of strategies and solution paths are presented in these problems.

Applications

The level of sophistication of the exercises increases from one set to the next. *Try It Out* presents problems in which basic understandings of the section concept are used. *Stretch Your Thinking* invites students to look beyond basic applications to a more sophisticated use of the concept. *Challenge Your Mind* presents fewer, more complex problems than the previous two applications. *Wrap It Up* is an application in which students either apply what they have learned throughout the section or use data to create a problem based on the focus of the section.

Assessment

There are multiple opportunities for assessment throughout **Problem Solving with Ratio and Proportion**. Each *Try It Out*, *Stretch Your Thinking*, and *Challenge Your Mind* page contains a suggestion for informal assessment. These are in the form of questions to pose to students in order to focus their attention on one important aspect of the problem-solving process.

Each *Wrap It Up* contains an assessment rubric that sets a performance standard for that activity. You may wish to review the rubric with students before they complete the *Wrap It Up*. This will let them know what is required to demonstrate a satisfactory level of understanding. *Wrap It Up* problems can become part of students' math portfolios.

How to Use **Problem Solving with Ratio and Proportion** in Your Classroom

Difficulty Level of the Problems

The word problems in this book have been written to appeal to a wide range of student abilities and interests. Within each level of application—*Try It Out*, *Stretch Your Thinking*, *Challenge Your Mind*—the level of analysis required increases. *Wrap It Up* activities are designed to be accessible to all students.

Using the Student Pages

Many of the problems can be approached by students working either independently, with a partner, or in a cooperative group. The *Solving Problems with…* problem is designed to be a teaching problem, presented as a teacher-led activity. You may also find that selected problems or sections are particularly appropriate at a given time, such as when a topic relates well to your daily math lessons.

- **Problem of the Day** If students possess the prerequisite skills needed for a particular section, the *Thinking About* problems are particularly suited to this approach. Have students solve one problem daily as a warm-up for any lesson.

- **Problem of the Week** More challenging problems from *Stretch Your Thinking* or *Challenge Your Mind* can be presented at the beginning of the week. You might post the problem on a class bulletin board. Students can work the problem during the week, asking you or other students for help as needed any time, up until the problem is due. Solutions can be discussed as a whole-class activity and perhaps posted on the bulletin board with the original problem.

- **Partners or Cooperative Groups** Students can work one or more problems in a section, arrive at solutions that group members agree upon, and present their solutions to the rest of the class. Disagreements over solutions can be worked out either within groups or as a whole class.

- **Whole-Class Lesson** In conjunction with studying a topic that is included in **Problem Solving with Ratio and Proportion**, you may wish to use a particular page as your math lesson for the day. Problems from the remainder of the page can be assigned as independent work or as a homework assignment.

Using the Teaching Pages

The main teaching technique in **Problem Solving with Ratio and Proportion** is the use of thought-provoking questions asked at the appropriate time. Much of the benefit to be gained from the program comes from students recognizing good questions. Over time, this will lead them to begin to ask themselves the kinds of questions needed to analyze a problem. However, this is a lengthy process and you should not look for results in the short run. A commitment to using the program and its approach to problem-solving will lead to success in the long run.

Problem-Solving Techniques

The Ten Solution Strategies

1. **Act Out or Use Objects** Acting out a problem helps you see the data and watch the solution process. You can also use objects to act out a problem.

2. **Make a Picture or Diagram** Drawing a picture or a diagram can help you see and understand the data in a problem.

3. **Use or Make a Table** Tables help you keep track of data and see patterns.

4. **Make an Organized List** Making a list makes it easier to review the data and helps you organize your thinking.

5. **Guess and Check** Guess the answer and check to see if it is correct. If your guess is incorrect, use what you learned to make more reasonable guesses until you solve the problem.

6. **Use or Look for a Pattern** Recognizing patterns can help you predict what will come next.

7. **Work Backward** Sometimes you can start with the data at the end of the problem and work backward to find the answer.

8. **Use Logical Reasoning** Use logical reasoning to solve problems that have conditions such as "If something is true, then…." Making a table often helps to solve logical reasoning problems.

9. **Make it Simpler** When a problem looks very hard, you can replace large numbers with smaller ones or reduce the number of items and solve that problem first.

10. **Brainstorm** When you cannot think of a strategy to use or a way to solve a problem, it's time to brainstorm. Stretch your thinking and be creative.

The Four-Step Plan

Step 1 Find Out	Find out what question the problem is asking you to answer, what information you have, and what information you need to get.
Step 2 Make a Plan	Make a plan that will help you solve the problem. Often, choosing a strategy is the first part of making a plan.
Step 3 Solve It	Solve it by using your plan to work through the problem until you find the answer to the question. If your plan doesn't work, try another approach.
Step 4 Look Back	Look back at the problem. Check that your solution answers the question and is a reasonable answer.

Problem-Solving Recording Sheet

Page: _____   Problem Number: _____

Find Out	○ What question must you answer to solve the problem?
	• What information do you have?
Make a Plan	○ How will you use the information you have to solve the problem?
Solve It	• Complete the solution, using the plan that you have made. Keep a record of your work.
	The Solution: _____
Look Back	○ Is your answer reasonable?
	• How would you check your solution?

1 | RATES
Introduction

This section focuses on rates and applications involving rates. A rate is a ratio that compares two quantities measured in different units. Unit rates, which are helpful in many consumer applications, will be considered fully in the next section.

Understanding Rate Problems

As students prepare to solve rate problems, they will need to consider how to write rates as ratios, which pieces of data to compare, and in which order the comparisons should be made. Consider the following problems:

Cereal is on sale at 2 boxes for $3.98. At that rate, how much will 7 boxes cost?
To solve this problem, students will need to:

- Recognize that $3.98 for 2 boxes is a rate, since it compares two quantities that use different units (money and boxes).

- Understand that the rate can be expressed as a ratio: $\frac{\$3.98}{2 \text{ boxes}}$.

- Decide on a solution method.

- Check that their results seem reasonable.

   Four friends went shopping for a class party. They found 3 containers of grape drink for $4.35. Will $12.00 be enough to buy 8 containers? How do you know?
   In a problem such as this, students must:

- Recognize that they are working with rates, although the term *rate* is not used.

- Determine what information is essential to solving the problem and what is not necessary. In this problem, the number of friends has no bearing on the solution.

- Understand that this is a multistep problem—they must first find the cost of 8 containers and then compare the cost to $12.

As students work with rate problems, ask these key questions as needed:

- What rate or rates do you see in the problem?

- How will using rates help you solve this problem?

- Do you see any proportions in the problem?

- Would you set up a proportion to solve this problem? Why or why not?

Solving Rate Problems

Your students will be more successful if they can combine general problem-solving strategies with methods specific to rate problems. Refer to *page vi* for a discussion of strategies.

Solving a Rate Problem Using a Proportion Write an appropriate proportion. Solve for the unknown.

Example: If it takes 90 min to input 6 pages, how long will it take to input 20 pages?

Set up a proportion as follows	$\frac{90 \text{ min}}{6 \text{ pages}} = \frac{x \text{ min}}{20 \text{ pages}}$
	$6(x) = 90(20)$
	$6x = 1{,}800$
	$x = 300$ min, or 5 hr

Solving a Rate Problem by Scaling Up Multiply both terms of the rate by the same factor or factors until you reach the needed value.

Example: A recipe calls for 5 tbs of applesauce for every 8 muffins. How much applesauce do you need for 48 muffins?

	Applesauce	Muffins	
Scale up as follows	5	8	
	15	24	Multiply both terms by 3.
	30	48	Multiply both terms by 2.

You need 30 tbs of applesauce for 48 muffins. Note that the numbers in any two rows of the table can be used to form a proportion.

Solving a Rate Problem by Scaling Down Divide both terms of the rate by the same divisor until you reach the needed value.

Example: An electric bill is $360 for 30 days. What is the average cost for 5 days?

	Days	Cost	
Scale down as follows	30	$360	
	5	$60	Divide both terms by 6.

The average cost of electricity for 5 days is $60.

Assessment

☑ **Informal Assessment** A suggestion for informal assessment will be found on each Try It Out, Stretch Your Thinking, and Challenge Your Mind page. The recommended question will help focus students' attention on one part of the problem-solving process.

Assessment Rubric An assessment rubric is provided for each Wrap It Up. Students' completed work may be added to their math portfolios.

Thinking About Rates

These problems will help students use their number sense and estimation skills as preparation for solving rate problems. Present one problem a day as a warm-up. You may choose to read the daily problem aloud, write it on the board, or create a transparency.

1. **A design on a banner uses 4 stars for every 6 circles. If the design is extended to contain 100 stars, will 120 circles be enough?**
   (No; 100 stars means there are 25 repetitions of the pattern, $100 \div 4 = 25$. For 120 circles, estimate that $6 \times 25 > 120$; so, 120 circles will not be enough.)

2. **A newspaper reported that 68 in. of snow fell last year. If the same amount falls this year, will there be at least 10 ft of snow over the 2 yr?**
   (Yes; estimate that 68 in. is more than 5 ft; so, 2 yr of more than 5 ft of snow is more than 10 ft of snow.)

3. **You and 4 friends are in line for the roller coaster ride. Together you have a total of $13. The sign at the booth reads "Tickets 2 for $6.50." Do you have enough money for all of you?**
   (No; four tickets cost twice $6.50 or $13; so, you don't have enough for five people.)

4. **If an airplane travels at an average speed of 620 mi/hr, can it make a 1,700-mi trip in 3 hr?**
   (Yes; estimate the product of 600×3; then compare $1,800 > 1,700$.)

5. **A family drove about 800 mi from New York City to Chicago in 16 hr. At that rate, can they make a 530-mi drive in 8 hr?**
   (No; if they drive 800 mi in 16 hr, then they would drive 400 mi in 8 hr—half the miles in half the hours.)

1 | RATES

Solving Problems with Rates

This lesson will help students focus on the kinds of questions they need to ask themselves in order to solve rate problems. Emphasize that there may be several approaches to solving a problem.

Using the Four-Step Method

Find Out

○ The problem asks students to find the number of extra math books at the middle school. Encourage students to describe the problem in their own words so that they understand that they are not simply being asked to find the total number of math books.

● Madison School District orders more math books than there are students.

○ The relationship between the number of math books and the number of students can be expressed in different ways:
a There are more math books than students.
b There are 12 math books for every 11 students.
If students use a proportion to solve the problem, the order in which they write the numbers in the rates depends on their understanding of the relationship of math books to students.

● Yes, students need to find the total number of math books bought in order to calculate the number of extra math books.

○ About 50 or 60 extra books; some students may estimate using the fact that there is 1 extra book for every 11 students. That's close to $\frac{1}{10}$. So, there should be about $\frac{1}{10}$ (or 0.1) of 550, or about 55 extra books.

Make a Plan

● Students will most likely scale up or use a proportion to find the total number of math books. They can then subtract to find the number of extra books.

Solve It

○ There are 600 – 550, or 50, extra math books.

These are two possible solution paths:

Scaling up	Students	Books	
	11	12	
	55	60	Multiply both terms by 5.
	550	600	Multiply both terms by 10.

Using a Proportion	
	$\frac{12 \text{ books}}{11 \text{ students}} = \frac{x \text{ books}}{550 \text{ students}}$
	$11x = 6{,}600$
	$x = 600$

Look Back

● Have students compare their results to their estimates. Ask: Is your solution reasonable when compared to your estimate? How did estimating help you solve the problem?

○ A way to check the solution is to solve by a different method. Encourage students to share their solution methods in small groups.

1 RATES

Solving Problems with Rates

Madison School District has over 3,000 students. To make sure each school has enough math books for all the students, the district buys 12 books for every 11 students. How many extra math books are there in the middle school if there are 550 students?

Find Out	○ What question must you answer to solve the problem?
	● What does the Madison School District do to make sure each school has enough math books?
	○ How are the numbers of math books and students related?
	● To solve the problem, do you need to know the total number of math books the school district buys? Why or why not?
	○ What is a reasonable estimate?
Make a Plan	● How can you use the relationship between the number of math books and the number of students to find the number of extra math books?
Solve It	○ Complete the solution using the plan that you have made. Keep a record of your work.
Look Back	● How does your solution compare to your estimate?
	○ How would you check to see whether your solution is correct?

1 | RATES

Try It Out

1. Find Out To help students understand the problem, ask:

- What does the problem ask you to find? (how many more gallons of gas are needed for a 500-mi trip)

- What ratios can you write to compare the amount of gas with the distance traveled? (gal/mi or mi/gal)

- How can you write a proportion that relates the amount of gas needed for a 350-mi trip to the amount needed for a 500-mi trip? ($\frac{14}{350} = \frac{x}{500}$)

- When you have solved the proportion, will you have the answer to the problem? Explain. (No, you must subtract 8 gal from the amount needed for the 500-mi trip.)

Solution Path

12 more gallons of gas are needed. $\frac{14}{350} = \frac{x}{500}$, $350x = 7,000, x = 20; 20 - 8 = 12$ gal

---

2. Yes; the first part of the trip is at a rate of 60 mi/hr, so they can travel the remaining 100 mi in less than 2 hr.

Some students may use the proportion $\frac{3.5}{210} = \frac{x}{100}$ to find that about 1.7 hr are needed to go 100 mi.

3. 20.4 mi; use a proportion, $\frac{0.17}{1} = \frac{x}{120}$

Encourage students to use estimation to check that their answers are reasonable.

---

4. $60.75; 3 groups of 3 is 9, so 3 × $20.25 is the answer

Some students may use a proportion:
$\frac{3}{\$20.25} = \frac{9}{x}, x = \$60.75.$
Some students may use the unit rate:
$20.75 ÷ 3 = \$6.75, \$6.75 × 9 = \$60.75.$

5. Yes, $5.00 is enough; 25($0.12) + 16($0.08) = $4.28

Help students see that the two rates, $0.12 per copy and $0.08 per copy, are part of the problem, but that the second rate applies only to copies beyond 25.

---

6. 12 c flour; the bigger recipe uses 4 times as many eggs, so it needs 4 times as much flour. 4 × 3 c = 12 c

☑ **Informal Assessment** Ask: What is the ratio of the number of eggs to the number of cups of flour in the original recipe? How does knowing that help you solve the problem? (5 eggs:3 cups; you can use that ratio to write a proportion)

7. About 1,050 words; $\frac{420}{2} × 5$

Some students may mentally calculate that there are 210 words per page and multiply that number by 5.

---

8. $1.74; (3 × $0.87) − $0.87

Students should recognize that what is asked for is the *difference* between 12 ears and 4 ears.

9. $111.60; use a proportion, $\frac{\$41.85}{3} = \frac{x}{8}$

Students may choose to find the cost of 1 CD and multiply by 8. Either way they are using the concept of proportionality to solve the problem.

---

1 | RATES

Try It Out

1. Your car uses 14 gal of gas to make a 350-mi round trip from Cleveland to Detroit. If you drive at the same average speed and already have 8 gal of gas in your car, how many more gallons will you need to make a 500-mi trip?

2. The Moreno family drove 210 mi between 8:00 a.m. and 11:30 a.m. They have another 100 mi to go. Can they complete their trip by 1:30 p.m.?

3. The length of a marathon race course is 26.2 mi. A world-class runner runs at a rate of 0.17 mi/min. What distance will the runner cover in the first 2 hr of the race?

4. Three friends pay a total of $20.25 to see a science-fiction movie. How much does the group of 9 friends in line behind them pay in all?

5. The copy shop charges $0.12 per copy for the first 25 copies and $0.08 per copy for any beyond 25. You have $5.00. Can you make 41 copies of a birthday party invitation?

6. A cake recipe uses 5 eggs for every 3 cups of flour. If you increase the recipe so that you use 20 eggs, how many cups of flour will you use?

7. Ernie's science report on Mars has 420 words in the first 2 pages. What is a reasonable estimate of the number of words in his report if it is 5 pages long?

8. The greengrocer sells 4 ears of corn for $0.87. At the same rate, how much more do 12 ears of corn cost than 4?

9. Three copies of this week's best-selling CD are $41.85. How much would 8 copies of the same CD cost?

1 | RATES

Stretch Your Thinking

1. Find Out To help students understand the problem, ask:

- What question does the problem ask you to answer? (How many songs will be played during a 3-hr program?)

   Make a Plan To help students decide on a strategy to follow, ask:

- If you knew the number of songs played in 1 hr, how could you solve the problem? (Multiply the number of songs played in 1 hr by 3.)

- If you knew the number of minutes in 3 hr, how could you solve the problem? (Possible answer: Solve the proportion $\frac{12 \text{ songs}}{45 \text{ min}} = \frac{x \text{ songs}}{180 \text{ min}}$.)

- If you knew the fractional part of an hour represented by 45 min, how could you solve the problem? (Possible answer: Solve the proportion $\frac{12 \text{ songs}}{0.75 \text{ hr}} = \frac{x \text{ songs}}{3 \text{ hr}}$.)

   Solution Paths
   48 songs will be played in a 3-hr program.

- $\frac{12 \text{ songs}}{45 \text{ min}} = \frac{x \text{ songs}}{60 \text{ min}}$, $x = 16$ songs in 1 hr, 16 songs in 1 hr $\times$ 3 hr = 48 songs; or

- $\frac{12 \text{ songs}}{45 \text{ min}} = \frac{x \text{ songs}}{180 \text{ min}}$; $x = 48$ songs; or

- $\frac{12 \text{ songs}}{0.75 \text{ hr}} = \frac{x \text{ songs}}{3 \text{ hr}}$; $x = 48$ songs

2. 0.62 mi/km; use a proportion, $\frac{93 \text{ mi}}{150 \text{ km}} = \frac{x \text{ mi}}{1 \text{ km}}$

Ask: Does the highway sign show a rate? Explain. (Yes, it shows a ratio of two quantities that are not measured in the same units.) Students can use the fact that a mile is longer than a kilometer to test their solutions for reasonableness.

3. a. 19,500 tickets per day; 58,500 ÷ 3

   b. 6,500 tickets per hr; 19,500 ÷ 3

4. 16 lemons

Students need to know that there are 32 oz in 1 qt in order to determine how many cans of ginger ale will be needed to make 3 qt.

3×32 oz = 96 oz
$96 \div 12 = 8$ cans
$2 \times 8 = 16$ lemons

Scaling up would be another reasonable solution method.

5. 70 additional revolutions;
$45 \times 6 = 270$, $33\frac{1}{3} \times 6 = 200$, $270 - 200 = 70$

Some students may find the difference for 1 min ($11\frac{2}{3}$ revolutions) and multiply that by 6. Answers should be rounded to the nearest whole number.

✓ **Informal Assessment** Ask: Which type of record has a faster rate? What does that mean? (Students should understand that 45 rpm is faster than $33\frac{1}{3}$ rpm and that it means that the smaller record turns faster than the larger one.)

1 | RATES

Stretch Your Thinking

1. The "Good Morning" radio program is on from 9 a.m. until noon every day. The radio station plays an average of 12 songs for every 45-min segment. At that rate, how many songs will be played during the "Good Morning" program?

---

2. In many places, highway signs give information in miles and kilometers. How many miles are equivalent to each kilometer?

> **Montreal**
> **93 mi**
> **150 km**

3. There were 58,500 tickets sold at ticket booths for a concert over 3 days. The booths were open each day from 6:00 p.m. until 9:00 p.m.

   a. What was the daily rate for ticket sales?

   b. What was the hourly rate for ticket sales?

---

4. To make your favorite fruit drink, you use 2 lemons for every 12-oz can of ginger ale. If you want to make 3 qt of the drink, how many lemons will you need?

5. Before CDs and cassettes were available, records were the most popular form of recorded music. Records of singles turned at 45 rpm (revolutions per minute) and LPs (long-playing records) turned at $33\frac{1}{3}$ rpm. For a 6-min song, how many additional revolutions did a 45 rpm record make than a $33\frac{1}{3}$ rpm record? Round your answer to the nearest whole number.

1 | RATES

Challenge Your Mind

1. **Find Out** To help students understand the problem, ask:

- What do parts **a** and **b** of the problem ask you to do? (Name rates that describe how much of the two types of food the cats eat.)

- What units will you compare to describe a rate for each food? (For the wet food, compare the number of cans to the number of days, weeks, or months. For the dry food, compare the number of ounces to the number of days, weeks, or months.)

- What will you need to consider to describe a rate for the amount of money Jen spends each month? (amounts of each type of food used, cost of each type, and number of units of time)

- Which type of food costs more? About how many times more expensive is the more costly food than the less costly? (The wet food costs more. It is about 3 times as expensive as the dry food.)

Solution Paths

a. Students may choose to describe the rate as an amount per day, per week, or per month. Per day: Since two 4-packs last 1 wk, 8 cans last 7 days. So, one possible rate is approximately 1 can per day. Other possibilities are: 8 cans per wk or 32 cans per mo or eight 4-packs per mo.

b. The dry food is used at a rate of 1 bag per mo or 48 oz per mo. Students might choose to assume that an average month has 4 wk or 30 days. If so, they might express the rate as $\frac{1}{4}$ bag per wk, 1 bag per 30 days, or 12 oz per wk, or 1.6 oz per day.

c. The monthly rate is $12.88.
Weekly rate: $\frac{(2 \times \$1.19) + (\$3.36 \div 4)}{1 \text{ wk}} = \$3.22$ per wk
The daily rate can be found by dividing:
$\$3.22 \div 7 = \$0.46$.
The monthly rate can be found by multiplying:
$\$3.22 \times 4 = \$12.88$.

2. a. 18 bussers; $\frac{3}{2} = \frac{x}{12}$
Some students may use scaling up.

b. 6 servers, 9 bussers
It may be helpful to review the *guess and check* strategy here.

c. $180 for servers; 6 servers $\times$ $5 per hr $\times$ 6 hr
$216 for bussers; 9 bussers $\times$ $4 per hr $\times$ 6 hr

✓ **Informal Assessment** Ask: Why does a ratio help you solve the problem? (It is a way to compare different units—servers and bussers are different units.)

1 | RATES

Challenge Your Mind

1. Jen has two cats. She feeds them a combination of wet and dry foods. She pays $1.19 for each 4-pack of wet food. She buys the dry food in 48-oz bags for $3.36. Two 4-packs of wet food last 1 wk. A bag of dry food lasts 1 mo.

a. Many different rates can be used to describe how much wet food Jen's cats eat. Name a rate that will help you find how much the cats eat in 1 mo. Explain.

b. Many different rates can be used to describe how much dry food Jen's cats eat. Name a rate that will help you find how much the cats eat in 1 mo. Explain.

c. Describe Jen's monthly expenses for the cats using a rate. You can assume that 1 mo is equal to 4 wk.

2. Sammy works in a restaurant where there are 3 bussers for every 2 servers. A server is paid $5 per hr. A busser is paid $4 per hr.

a. When the owner of the restaurant expects a large crowd, he has 12 servers working at the restaurant. How many bussers are needed for 12 servers?

One night, the restaurant was open from 5:00 p.m. until 11:00 p.m. After closing, the owner paid a total of $396 in salary to all the servers and bussers. There were 15 of them in all.

b. How many servers worked that night? How many bussers?

c. How much of the salary went to the servers? How much went to the bussers?

1 | RATES

Wrap It Up

Take a Break!

You might wish to discuss the problem with the class before they begin to work on their own. Ask:

- How can you use the information in the problem to find the number of drinks sold? (Subtract to find the number of each drink sold, then add the results.)

- If you know how many drinks were sold during 2 wk, how can you find the total amount earned by the machine? (Multiply the total by $0.75.)

- If you know the number of drinks sold during a 2-wk period, how can you find the number sold each week? (Divide by 2.)

- If you write the rate per 2 wk and the rate per week as fractions, what should be true about the fractions? (They should be equivalent.)

Solutions

a. The machine earns approximately $64.50 per wk; $12.90 per day.

b.

Beverage	Per Week	Per Day
Water	22 bottles	about 4 bottles
Apple juice	28 cans	about 6 cans
Grape juice	20 cans	about 4 cans
Orange drink	16 cans	about 3 cans

c. Apple juice is selling at the fastest rate.

Possible Solution Paths

a. $64.50 per wk; $12.90 per day.
220 (to start) – 48 (left) = 172 drinks sold
172 × $0.75 = $129
$129 ÷ 2 = $64.50 per wk
$129 ÷ 10 = $12.90 per day

b.

Beverage	Per Week	Per Day
Water	$(50 - 6) \div 2 = 22$ bottles	$(50 - 6) \div 10 \approx 4$ bottles
Apple juice	$(70 - 14) \div 2 = 28$ cans	$(70 - 14) \div 10 \approx 6$ cans
Grape juice	$(60 - 20) \div 2 = 20$ cans	$(60 - 20) \div 10 \approx 4$ cans
Orange drink	$(40 - 8) \div 2 = 16$ cans	$(40 - 8) \div 10 \approx 3$ cans

c. Per week: 28 > 22 > 20 > 16
Compare the rates. The drink with the greatest weekly or daily rate is selling fastest.

Assessment Rubric

3 The student correctly determines the number of each type of drink sold; accurately describes the amount of money earned by the machine using a rate comparing money to a unit of time; finds the average weekly rate and daily rate for each drink, rounding correctly where needed; and correctly determines the fastest-selling drink.

2 The student correctly determines the number of each type of drink sold; determines the amount of money earned by the machine, but may have difficulty associating the amount with a unit of time; has some success determining the average weekly and daily rates for at least several of the drinks; and makes a reasonable guess at the fastest-selling drink.

1 The student correctly determines the number of each type of drink sold, but has difficulty relating the number to the amount of money earned by the machine; has limited success in determining a rate of sale, whether weekly or daily; and cannot determine which drink is selling at the fastest rate.

0 The student may be able to determine the number of each type of drink sold, but does not show any understanding of rate.

1 | RATES

Wrap It Up

Take a Break!

A school beverage machine is filled with 50 bottles of water, 70 cans of apple juice, 60 cans of grape juice, and 40 cans of orange drink. After 2 wk of school (10 days), there are 6 bottles of water, 14 cans of apple juice, 20 cans of grape juice, and 8 cans of orange drink left.

a. It costs $0.75 to buy a drink. What is the average amount the machine earns per week? Per day? Write your answers as rates.

b. On average, how many bottles or cans of each drink are selling per week? Per day? If necessary, round each rate to the nearest whole number.

c. Which drink is selling at the fastest rate?

---

2 | UNIT RATES
Introduction

The focus of this section is problem solving with unit rates. Unit rates are helpful in comparing quantities such as costs, car mileage, salaries, or speed. A unit rate is a rate that compares a quantity to 1 unit. Examples of unit rates are $1.99 per lb and 45 words per min. When unit rates are written as fractions, the number 1 is usually omitted before the second term (65 mi/hr instead of $\frac{65 \text{ mi}}{1 \text{ hr}}$).

Understanding Unit Rate Problems

To be successful with the problems in this section, students need an understanding of ratio and proportion, as well as the meaning of rate.

Using unit rates in everyday life requires a little common sense. Unit rates are usually best for comparing things of the *same* category. Decisions about which of two items to buy can be made on a mathematical basis, but personal preference may also play a role. Have students compare the following problems:

A 25-oz box of cereal costs $3.99. An 18-oz box of the same cereal costs $2.99. Which box is the better buy?

A 2-lb steak at the supermarket costs $6.98. Three pounds of chicken cost $4.47. Which is the better buy?

- The first problem can be solved by finding the unit price of each size box. All other things being equal, the better buy is the box with the lower unit cost (that is, the lower price per ounce).

- The second problem can be examined by comparing the price per pound of the two items. A customer might, however, prefer steak to chicken and decide to buy the steak even though it costs more per pound.

Solving Unit Rate Problems

Unit rate problems can be solved using the same strategies as other rate problems. Refer to page *vi* for a discussion of problem-solving strategies. Since the second term in a unit rate is always 1 unit, problems can often be solved by simple division.

Solving a Unit Rate Problem Using a Proportion

Proportional thinking, number sense, and estimation skills are important when using unit rates to solve problems and evaluate solutions for reasonableness.

Example: On a test, a word processor was able to correctly input 141 words in 3 min. What was his rate?

Write and solve the following proportion	$\dfrac{141 \text{ words}}{3 \text{ min}} = \dfrac{x \text{ words}}{\text{min}}$

Solving the proportion gives an answer of 47 words per min. Since 141 ÷ 3 is approximately 150 ÷ 3, or 50, an answer of 47 words per min is reasonable.

Solving a Unit Rate Problem Using Scaling Problems containing compatible numbers often can be easily solved by scaling down.

Example: If a car is able to travel 240 mi on 8 gal of gasoline, how many miles does it get per gallon?

	Distance traveled	Gasoline used
Scale down as follows	240 mi	8 gal
	120 mi	4 gal
	60 mi	2 gal
	30 mi	1 gal

The car averages 30 mi/gal. Note that miles per gallon is a unit rate.

Solving a Unit Rate Problem by Division Since finding a unit rate results in one of the terms being 1 unit, unit rate problems are often easily solved by division.

Example: If a dozen bagels costs $5.40, what is the unit cost?

Once students understand that the problem asks for the cost of a single bagel or 1 unit, they can find the solution by division: $5.40 \div 12 = \$0.45$.

Assessment

✓ **Informal Assessment** A suggestion for informal assessment will be found on each Try It Out, Stretch Your Thinking, and Challenge Your Mind page. The recommended question will help focus students' attention on one part of the problem-solving process.

Assessment Rubric An assessment rubric is provided for each Wrap It Up. Students' completed work may be added to their math portfolios.

Thinking About Unit Rates

These problems help students use their number sense and estimation skills as preparation for solving unit rate problems. Present one problem a day as a warm-up. You may choose to read the daily problem aloud, write it on the board, or create a transparency.

1. **Anthony measured his height and estimated he had grown 9 cm in 2 mo. Is his estimate reasonable? If not, what might he have done incorrectly?**
   (The estimate is not reasonable. He probably mistook millimeters for centimeters.)

2. **A 6-oz can of frozen juice is mixed with 3 cans of water. Does the can make at least one quart of juice? How do you know?**
   (No; 6 oz of frozen juice + 3 × 6 oz of water = 24 oz, which is less than 32 oz or 1 qt.)

3. **Eggs are $1.09 per dozen. An omelet uses 3 eggs. Is $5.00 enough to buy eggs for 12 omelets?**
   (Yes; 4 omelets from a dozen eggs, 12 omelets from 3 dozen eggs; 3 × $1.09 < $5.00.)

4. **A diet drink has 1 calorie per can. If there are 2 servings in the can, how many servings are there in a 6-pack of the drink? How many calories?**
   (12 servings; 6 calories)

5. **A bakery special is offering 3 muffins for $0.96. Can Emily buy a half dozen muffins with $1.50?**
   (No; 3 muffins cost about $1.00; so, 6 muffins cost about $2.00. Or muffins cost about $0.30 each, 6 × $0.30 = $1.80.)

2 | UNIT RATES

Solving Problems with Unit Rates

This lesson extends students' understanding of rates to unit rates. It will help students see how using unit rates enables them to make comparisons.

Using the Four-Step Method

Find Out
- ○ The problem asks students to determine which runner ran at a faster rate.
- ● It took Wilson Kipketer 101.11 sec. It took Hicham El Guerrouj 206.00 sec ($3 \text{ min} \times 60 \frac{\text{sec}}{\text{min}} + 26 \text{ sec} = 206 \text{ sec}$). Point out that since the distances for the races were different, the shorter time may not indicate a faster rate.
- ○ In order to decide which runner was faster, students can calculate the number of meters per second for each runner. Many students are likely to do this, since it is analogous to miles per hour for automobiles. Students can also use the number of seconds per meter to decide who was faster.
- ● The faster runner is the runner that covers a greater distance in 1 unit of time. If the student chooses to use seconds per meter for the unit rates, then the faster runner is the one who takes less time to run 1 m.

Make a Plan
- ○ Before students actually solve the problem, have volunteers describe to the class what they are planning to do and why.

Solve It
- ● Kipketer had a faster rate.

  Using seconds per meter, students can divide to find the unit rates.

  Kipketer: $\frac{101.11 \text{ sec}}{800 \text{ m}} = \frac{0.13 \text{ sec}}{\text{m}}$

  El Guerrouj: $\frac{206.00 \text{ sec}}{1,500 \text{ m}} = \frac{0.14 \text{ sec}}{\text{m}}$

  Using meters per second:

  Kipketer: $\frac{800 \text{ m}}{101.11 \text{ sec}} = \frac{8 \text{ m}}{\text{sec}}$

  El Guerrouj: $\frac{1,500 \text{ m}}{206.00 \text{ sec}} = \frac{7 \text{ m}}{\text{sec}}$

Look Back
- ○ Number sense can help students check that answers are reasonable. For example, if Kipketer had doubled both his time and distance, he would have covered 1,600 m in almost the same time it took El Guerrouj to cover 1,500 m.
- ● Some students may recognize that by dividing Kipketer's distance and time by 8 and El Guerrouj's distance and time by 15, they will have a comparison in which they have a time for 100 m for each runner. Encourage students to share their solution methods in small groups.

2 | UNIT RATES

Solving Problems with Unit Rates

In 1997, Wilson Kipketer set a world record in the men's 800-m race with a time of 1 min 41.11 sec, or 101.11 sec. In 1998, Hicham El Guerrouj set a world record in the men's 1,500-m race with a time of 3 min 26.00 sec. Which runner ran at a faster rate?

Find Out	○ What question must you answer to solve the problem?
	● How long did it take each runner to complete his race?
	○ What units can you use in order to compare their speeds?
	● When you compare unit rates, how will you decide whether the faster runner has a greater or lesser unit rate?
Make a Plan	○ Once you understand the rates that you are comparing, how will you use that information to solve the problem?
Solve It	● Complete the solution using the plan that you have made. Keep a record of your work.
Look Back	○ How will you check that your answer is reasonable?
	● Can you think of another way that you could have solved the problem?

2 | UNIT RATES
Try It Out

1. Find Out To help students understand the problem, ask:

- What does the problem ask you to find? (the amount the school saves by buying 3 cartons of cereal rather than individual boxes)

- Which unit prices do you already know? Which must you find? (price per box bought individually, price per carton; price per box from the carton)

- Which unit is most reasonable to use for comparison? Why? (one box of cereal, since the unit price of an individual box is given and the unit price of a box from the carton can be found)

Solution Path

$18.00; $29.90 ÷ 10 = $2.99 per box bought by the carton
$3.59 − $2.99 = $0.60 savings per box
30 × $0.60 = $18.00 is the total savings for your school

---

2. 16 hr; use a unit rate of $\frac{1}{16}$ hall per hr to determine that it takes a total of 16 hr to paint the hall

Students may assume, incorrectly, that if 2 painters take 8 hr, 1 painter would need 4 hr ($\frac{2}{8} = \frac{1}{4}$). Be sure students apply the test of reasonableness to their work.

3. 26-oz bag for $3.38; $3.38 ÷ 26 oz = $0.13 per oz, $2.24 ÷ 16 oz = $0.14 per oz

Discuss students' ideas for checking their solutions.

---

4. 504 mi; use a unit rate, 441 mi per 7 hr = 63 mi/hr, 8 × 63 = 504 mi

Some students may use a proportion to solve:

$$\frac{441 \text{ mi}}{7 \text{ hr}} = \frac{x}{8 \text{ hr}}$$
$$x = 504 \text{ mi}$$

5. $17.40

Students need to know that there are 4 qt in 1 gal.

$3.29 ÷ 2 = $1.645 per qt (jug)
$1.79 − $1.645 = $0.145 savings per qt
30 gal = 120 qt
120 × $0.145 = $17.40 total savings

---

6. No, they will each ride 9 mi in 45 min (12 mi/hr × $\frac{3}{4}$ hr). To pass each other, they have to reach the halfway point of 10 mi.

Students may find the *act out* or *make a diagram* strategies helpful in solving this problem.

7. About 130 mi/hr; 2,477.8 ÷ 19.083 ≈ 130 mi/hr

☑ **Informal Assessment** Ask: When you write a unit rate for this problem, what value will you make the unit? Why? (Students need to consider whether to use minutes or hours. The more reasonable unit to use is hours.)

---

8. $3.66; (10 × $1.39) − (3 × $2.95 + $1.39).

9. 12 mi/gal more; $\frac{640}{20} − \frac{400}{20} = 12$ mi/gal

Encourage students to use mental math to solve this problem.

Name _____

2 | UNIT RATES
Try It Out

1. A 10-box carton of cornflakes costs $29.90. How much would your school save by buying 3 cartons instead of the same number of boxes, each priced at $3.59?

2. Two painters working at the same pace need 8 hr to paint a hall. How long would it take one of them to do the job alone?

3. Which is the better buy: a 26-oz bag of Crispy Crackers for $3.38 or a 1-lb bag of the same crackers for $2.24?

4. You have driven 441 mi in 7 hr. If you continue to drive at this rate, how far will you drive in 8 hr?

5. A quart carton of orange juice costs $1.79 and a half-gallon jug costs $3.29. You expect to use 30 gal during the year. What will your yearly savings be if you buy the half-gallon jugs?

6. Tina and Lisa live on the same road, 20 mi apart. Tina leaves her home and rides toward Lisa's house at a rate of 12 mi/hr. At the same time, Lisa leaves her house and rides toward Tina's house at the same rate. Will they pass each other within 45 min? Explain.

7. On August 24, 1932, Amelia Earhart set a women's speed record by flying 2,477.8 mi from Los Angeles to Newark in 19 hr 5 min. What was her average rate of speed to the nearest mile?

8. How much do you save if you buy 10 markers the least expensive way rather than all individually?

9. A car with a full 20-gal tank of gas can travel 640 mi on the highway. In the city, the car can travel 400 mi with a full tank. How many more miles per gallon does the car get on the highway than in the city?

© Creative Publications 10874

Problem Solving with Ratio and Proportion 19

2 | UNIT RATES

Stretch Your Thinking

1. Find Out To help students understand the problem, ask:

- What does the problem ask you to find? (which job pays better; which job will enable Cairy to earn $241.50 more quickly)

  Make a Plan To help students decide on a strategy to follow, ask:

- What does "better-paying job" mean? (the job with the greater hourly rate)

- How can you decide which is the better-paying job? (Find the hourly rate for each job and compare.)

- Will the "better-paying job" be the one that meets Cairy's needs? (Not necessarily; the job with the greater hourly rate may not earn her the amount she needs more quickly.)

Solution Path

Cairy should choose the balloon-delivery job; it pays less per hour but more per week, so she will earn enough sooner.

$47.25 ÷ 9 = $5.25 and $38.50 ÷ 7 = $5.50; the fast-food job is the better-paying job.

$241.50 ÷ $47.25 ≈ 5.1 wk and $241.50 ÷ $38.50 ≈ 6.3 wk; 5.1 wk < 6.3 wk.

2. 2.4 oz; 96 ÷ 2 = 48 oz, 48 ÷ 20 = 2.4 oz per baguette

☑ **Informal Assessment** Ask: How would you explain to someone who has been absent how you chose the unit to use? (Students' explanations should include the fact that the units for the two parts of the rate must be different, that one of the parts must be 1, and that the unit rate should accurately reflect the information in the problem.)

3. Yes, the label is incorrect; $3.45 ÷ 72 ≈ $0.048 per ounce. To correct the label, round the unit cost to the nearest cent as $0.05 per oz or to the nearest thousandth as $0.048 per oz.

You may wish to allow calculators for calculating the unit price.

4. About 26 yr old; use a proportion,

$$\frac{3}{115,185,000} = \frac{x}{1,000,000,000}$$

Some students may estimate heartbeats for 1 year and then use a proportion to find Rosa's age at 1 billion heartbeats. Discuss any other solution strategies that students suggest.

You may wish to allow calculators for calculating Rosa's age.

5. 0.025 gal per mi or 40 mi per gal;

$$\frac{3\text{ gal}}{120\text{ mi}} = \frac{(3 ÷ 120)\text{ gal}}{\text{mi}}$$

Before beginning to solve the problem, it may be necessary to have several students read the problem aloud. Ask: What kind of unit rate is most appropriate for the problem? (Students need to find the number of gallons needed to travel 1 mi, not the more typical number of miles driven on 1 gal.)

2 | UNIT RATES

Stretch Your Thinking

1. Cairy can earn $47.25 for 9 hr of work per week working for a balloon-delivery service. She can earn $38.50 for 7 hr of work per wk as a server in a fast-food shop. Cairy needs $241.50 for a mountain bike and would like to earn the money as fast as possible. Which is the better-paying job? Which job should she take to accomplish her goal?

2. A baker has 6 lb, or 96 oz, of bread dough. He uses 3 lb of the dough to make baguettes. He makes 20 baguettes, each with the same weight. How many ounces does each baguette weigh before it is baked?

3. In the supermarket where you work after school, six 12-oz cans of soda cost $3.45. A label identifies the unit cost as $0.047 per oz. Is the label incorrect? Explain how you would correct the label if there is an error.

4. In the first three years of Rosa's life, her heart beat about 115,185,000 times. About how old will she be when it beats its billionth time? (Hint: An average year is 365.25 days.)

5. A small car can travel 120 mi on 3 gal of gas. At what rate does the car use gasoline?

Problem Solving with Ratio and Proportion

2 | UNIT RATES
Challenge Your Mind

1. Find Out To help students understand the problem, ask:

- What questions does the problem ask you to answer? (on which ride will the students finish first and by how many minutes)

- What unit rates are described in the problem? (people per line, time per ride, people per ride, reloading time per ride)

- What do you need to consider to find the total time for each group on each ride? (the length of each ride and the time between rides)

Solution Path
The group on Flyaway will finish 2.5 min before the group on Speed Bumps.

Find the number of rides needed for the people in each line.
Speed Bumps: 4 rides, since $93 \div 25 = 3$ R18;
Flyaway: 3 rides, since $87 \div 40 = 2$ R7

Find the total times of the rides.
Speed Bumps: $4 \times 2.5 = 10$ min;
Flyaway: $3 \times 3.5 = 10.5$ min

Find the times to load groups and to unload the final group.
Speed Bumps: $5 \times 3 = 15$ min;
Flyaway: $4 \times 3 = 12$ min

Find the total times.
Speed Bumps: $10 + 15 = 25$ min;
Flyaway: $10.5 + 12 = 22.5$
25 min $- 22.5$ min $= 2.5$ min

2. Students can make use of the *guess and check* strategy in combination with the *make a table* strategy to solve this problem. If your class has studied graphing functions, parts **a** and **b** can be solved graphically.

a. Doreen should choose CallPro. The cost on American Phone Lines would be $19.99 + (120 - 60)$0.40 = 43.99. The cost on CallPro would be $29.99, since 120 min is less than 180 min. $43.99 > $29.99.

b. For 85 min/mo, the cost for the two companies is the same, $29.99.
$29.99 - $19.99 = $10.00; $10.00 \div $0.40 = 25;
$25 + 60 = 85$ min

c. For fewer than 85 min/mo, choose American Phone Lines.
For 85 min/mo, choose either company.
For more than 85 min/mo, choose CallPro.

✓ **Informal Assessment** Ask: Why do you think CallPro can charge so much more for extra minutes than American Phone Lines? (Although CallPro charges a higher base rate, 3 times as many minutes are included in the rate.) Students should see that consumers often make decisions based on comparison shopping.

2 | UNIT RATES
Challenge Your Mind

1. One Saturday morning, a school group of 180 students visits the amusement park. Ninety-three students get in line for Speed Bumps, the rest get in line for Flyaway. A ride on Speed Bumps lasts 2 min 30 sec. A ride on Flyaway lasts exactly 1 min longer. It takes 3 min for everyone to get on a ride or to get off while the next group is getting on.

   If 25 people can ride Speed Bumps at one time and 40 people can ride Flyaway, which group of students will be finished first? How many minutes before the other group will they be finished?

2. Doreen uses a cell phone about 2 hr each month. She is deciding between these two new cellular phone companies.

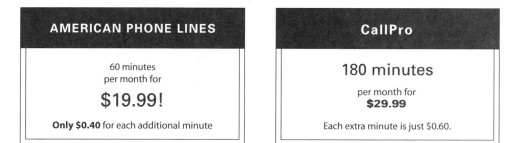

AMERICAN PHONE LINES	CallPro
60 minutes per month for	180 minutes
$19.99!	per month for **$29.99**
Only **$0.40** for each additional minute	Each extra minute is just $0.60.

a. Which company is probably a better buy for Doreen? Explain why you think so.

b. For how many minutes of monthly use would the cost of using the two companies be exactly the same?

c. What rule could you write to help someone make the best choice between the two phone companies?

2 | UNIT RATES
Wrap It Up

World Record Rates

Part 1 Discuss how to interpret the times in the table.

- For the men's 5,000-m race, what does the time 12:39.36 mean? (12 min 39.36 sec)

- How many seconds is 12 min 39.36 sec, rounded to the nearest second? (759 sec)

- Why would you use the same type of unit rate for all the races, regardless of the length of the race? (Students should understand that if they are going to compare rates, they should all be given in the same units.)

- Which unit makes the most sense? (Answers may vary.)

Solution

The rates for the record times (rounded) are:

Men's 300 m: 9.5 m/sec
Men's 400 m: 9.3 m/sec
Men's 1,000 m: 7.6 m/sec
Men's 3,000 m: 6.8 m/sec
Men's 5,000 m: 6.6 m/sec

Women's 200 m: 9.4 m/sec
Women's 800 m: 7.1 m/sec
Women's 1,500 m: 6.5 m/sec
Women's 3,000 m: 6.2 m/sec
Women's 5,000 m: 5.8 m/sec

As the distances increase, the unit rates decrease for both men and women.

Students will need to compute to tenths to discern differences in rates.

Part 2 Information about current world records can be found in an almanac.

Ask students:

- When you compare unit rates over time for the races, which rates do you think will have changed the most? The least? (Answers may vary.)

Assessment Rubric

3 The student correctly finds a unit rate for each World Track record time using the same units, accurately describes the pattern in the resulting data, compares rates, finds the rates for a recent Olympics, and describes changes in the rates over time.

2 The student correctly names a unit to use for each race and correctly finds some unit rates, describes the pattern in the data, compares rates, finds the rates for a recent Olympics, and shows some understanding of changes that have taken place over time.

1 The student has difficulty selecting a unit to use for each race, with help finds unit rates for at least one of the races but cannot describe the pattern in the data, finds data from a recent Olympics but cannot describe differences over time.

0 The student cannot apply the concept of unit rates, does not see the pattern in the data, may find recent Olympics data but cannot apply them to the task of describing change over time.

2 | UNIT RATES
Wrap It Up

World Record Rates

The table shows past outdoor world records for several men's and women's track events.

Part 1

How will you interpret the times in the table? Find a unit rate to describe each of the winning times. What do you notice about the unit rates as the distances of the races increase?

Part 2

a. In the 1964 Olympics, the winning time in the men's 400-m race was 45.1 sec. The winning time in the women's 200-m race was 23.0 sec. Compare these rates to the current world record rates.

b. Choose any of the races above. Predict and then find the winning time for that race in a recent Olympics. What has happened to the rate you used in part 1? Was your prediction accurate?

World Track Records	
Event	Time (min:sec)
Men's 300-m race	31.48
Men's 400-m race	43.18
Men's 1,000-m race	2:11.96
Men's 3,000-m race	7:20.67
Men's 5,000-m race	12:39.36
Women's 200-m race	21.34
Women's 800-m race	1:53.28
Women's 1,500-m race	3:50.46
Women's 3,000-m race	8:06.11
Women's 5,000-m race	14:28.09

3 ENLARGEMENTS & REDUCTIONS

Introduction

This section focuses on enlargements and reductions, which are an application of similar figures. Two similar figures have exactly the same shape, though possibly different sizes. If the sizes are the same, the figures are congruent. If the sizes are different, the larger figure is an enlargement of the smaller and the smaller figure is a reduction of the larger.

Understanding Enlargement and Reduction Problems

In similar figures, corresponding angles must be congruent. Lengths of corresponding sides must have been multiplied or divided by the same factor to ensure that the sides are in proportion.

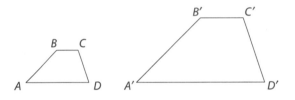

Two similar trapezoids can help students make comparisons such as:

- Corresponding angles are angles A and A', B and B', C and C', D and D'.

- The corresponding angles are the same size.

- For similar figures, corresponding sides are related by the same scale factor. The scale factor that was used to enlarge trapezoid $ABCD$ is 2.

- Ratios of corresponding sides between figures are the same in similar figures. An example of a ratio for these trapezoids is $\frac{AB}{AD} = \frac{A'B'}{A'D'}$.

Solving Enlargement and Reduction Problems

Your students will be more successful if they can combine general problem-solving strategies with methods specific to enlargement and reduction problems. Some strategies that may be especially useful are *make a picture or diagram*, *use or make a table* (for scaling up and scaling down), *guess and check*, and *make it simpler*. Refer to page *vi* for a discussion of strategies.

Enlargement and reduction problems can be solved using the same strategies as other proportion problems. Proportional thinking, number sense, and estimation skills are important to solve problems and evaluate solutions for reasonableness.

Solving Enlargement and Reduction Problems by Using Proportions Proportions are a way to solve enlargement and reduction problems.

Example: Use the trapezoids at left. Suppose that $AB = 6$ cm, $CD = 4.5$ cm, and the corresponding sides in trapezoid $A'B'C'D'$ are twice as long. What equivalent ratios can be written about the trapezoids?

Trapezoid $A'B'C'D'$ is an enlargement of trapezoid $ABCD$ with a scale factor of 2. Equivalent ratios can be set up in either of the following ways:

$$\frac{AB}{CD} = \frac{A'B'}{C'D'} \text{ and } \frac{AB}{A'B'} = \frac{CD}{C'D'}$$

$$\text{since } \frac{6 \text{ cm}}{4.5 \text{ cm}} = \frac{12 \text{ cm}}{9 \text{ cm}} \text{ and } \frac{6 \text{ cm}}{12 \text{ cm}} = \frac{4.5 \text{ cm}}{9 \text{ cm}}$$

Reversing the numerator and denominator in each ratio also results in a correct proportion.

Solving Recipe Problems Recipes are often enlarged or reduced. While there are no linear measurements involved, the amount of each ingredient must also be multiplied or divided by the same factor. Scaling up or scaling down is often a useful method to find changes in recipe amounts.

Example: Suppose you want to make a recipe that uses 3 c of carrots and 2 c of rice but you have only 1 c of carrots. How much rice should you use?

Since you have $\frac{1}{3}$ the carrots for the full recipe, you can make $\frac{1}{3}$ the recipe. Multiply the amount of rice by $\frac{1}{3}$. You will need $\frac{2}{3}$ c rice for the reduced recipe.

As always, encourage students to check that their solutions make sense.

Assessment

✓ **Informal Assessment** A suggestion for informal assessment will be found on each Try It Out, Stretch Your Thinking, and Challenge Your Mind page. The recommended questions will help focus students' attention on one part of the problem-solving process.

Assessment Rubric An assessment rubric is provided for each Wrap It Up. Students' completed work may be added to their math portfolios.

Thinking About Enlargements and Reductions

These problems help students use their number sense and estimation skills as preparation for solving enlargement and reduction problems. Present one problem a day as a warm-up. You may choose to read the daily problem aloud, write it on the board, or create a transparency.

1. **Draw a rectangle (not a square) on the board. Ask: What is the least number of these tiles that you would need to make a larger tile with the same shape?**
(4; using 2 tiles would double either the length or width, but not both.)

2. **Draw a unit square with sides 1 unit long. Then draw a square with sides 2 units long. Compare their areas.**
(The area of the unit square is 1 square unit; the second square is 4 square units. Although the lengths of the sides are doubled, the area is quadrupled.)

3. **Is an adult an enlargement of the person as a baby? Why or why not?**
(No; the ratios of body dimensions change.)

4. **Draw on the board:**

Ask students: In this drawing, the father is taller than his son. The father is 6 ft tall. How could you use the drawing to estimate the height of his son?
(You could measure the heights of the stick figures, then set up and solve a proportion using the measurements and the known height.)

5. **A fruit punch recipe that serves 4 people uses 3 c of orange juice. If you have 10 c of orange juice, can you make punch for 16 people?**
(No; for 16 people, you would need 12 c.)

3 | ENLARGEMENTS & REDUCTIONS
Solving Problems with Enlargements and Reductions

This lesson will help students focus their attention on the idea that an enlargement or reduction creates a pair of similar figures. They will investigate the connections that need to be made between the original and the resulting figure.

Using the Four-Step Method

Find Out	○ Students are asked to find the dimensions of the largest copy of Kinsung's drawing that will fit on the piece of cardboard.
	● Some students may suggest using proportions. Others may prefer to try various scale factors to see if they can enlarge or reduce one of the rectangles to obtain the other rectangle.
	○ Students can use the fact that 5:18 is between 3 and 4 and 4:15 is between 3 and 4. So the scale factor will be between 3 and 4. This means that the picture will be smaller than 5 in. × 6 in. but larger than $3\frac{3}{4}$ in. x $4\frac{1}{2}$ in.
Make a Plan	● Students can use their knowledge of multiplication of fractions to see what scale factors would allow them to reduce a side of the larger rectangle to exactly fit one of the sides of the smaller rectangle. Scale factors worth checking are $\frac{5}{15}, \frac{5}{18}, \frac{4}{15}$, and $\frac{4}{18}$.
	○ Students will probably realize that the only scale factors they need consider are those that change the dimensions of the larger rectangle to dimensions that are equal to or less than those of the smaller rectangle.
	● Students can reason that the reduced rectangle that fits on the cardboard and has the correct dimensions gives the solution.
Solve It	○ The rectangle that is 4 in. by $4\frac{4}{5}$ in. is the largest reduction that works. $15 \times \frac{5}{15} = 5$ and $18 \times \frac{5}{15} = 6$ (the enlargement is too big) $15 \times \frac{5}{18} = 4\frac{1}{6}$ and $18 \times \frac{5}{18} = 5$ (the enlargement is too big) $15 \times \frac{4}{15} = 4$ and $18 \times \frac{4}{15} = 4\frac{4}{5}$ (the enlargement will fit) $15 \times \frac{4}{18} = 3\frac{1}{3}$ and $18 \times \frac{4}{18} = 4$ (the enlargement will fit)
Look Back	● Ask students whether they think their solution is close enough to their estimate for them to accept. Since the 4 × 5 piece of cardboard is not proportional to the original drawing, only one dimension can be fully utilized. This gives the largest possible copy rather than a copy that is proportional to the drawing and fits on the 4 in. × 5 in. piece of cardboard.

3 ENLARGEMENTS & REDUCTIONS

Solving Problems with Enlargements and Reductions

Kinsung has a drawing that is 18 in. long and 15 in. tall. He wants to mount a copy of the drawing on a piece of cardboard that is 4 in. by 5 in. What are the dimensions of the largest copy he can use?

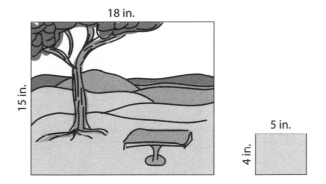

Find Out	○ What question must you answer to solve the problem?
	● How can you decide whether the drawing is similar to the cardboard rectangle?
	○ Estimate the dimensions of the largest copy that will work.
Make a Plan	● How will you decide what scale factors to consider?
	○ How can you tell whether a particular scale factor will give you a copy that fits on the cardboard?
	● How can you tell when you have found the largest copy that will work?
Solve it	○ Complete the solution using the plan that you have made. Keep a record of your work.
Look Back	● How does your solution compare to your estimate?

3 ENLARGEMENTS & REDUCTIONS

Try It Out

If possible, display some drawings or photographs and a 75% reduction of each. Discuss how measuring and comparing lengths can be used to find the reduction factor.

1. Find Out To help students understand the problem, ask:

- What does the problem ask you to find? (how many inches shorter the car in the copy will be than the original)

- How will the length of any part of the reduced drawing compare to the corresponding part of the original? (It will be 75% or $\frac{3}{4}$ the length of the original.)

- How will the shape of the car in the reduction compare to its shape in the original drawing? (The shapes will be the same; the figures will be similar.)

Solution Paths

- 2.25 in. Students can set up and solve a proportion. For 75%, use the ratio $\frac{75}{100}$ or $\frac{3}{4}$.

$$\frac{x}{9} = \frac{75}{100} \qquad \frac{x}{9} = \frac{3}{4}$$
$$100x = 75(9) \qquad 4x = 3(9)$$
$$100x = 675 \qquad 4x = 27$$
$$x = 6.75 \text{ in.} \qquad x = 6.75 \text{ in.}$$

Subtract to get the answer: $9 - 6.75 = 2.25$ in.

- Students may also use the scale factor to solve. The scale factor is 0.75 or $\frac{3}{4}$. Multiply $9 \times \frac{3}{4} = \frac{27}{4}$, or $6\frac{3}{4}$ in. Then subtract $9 - 6\frac{3}{4} = 2\frac{1}{4}$ in.

2. $2\frac{1}{2}$ c; use a unit rate to solve:
$1\frac{1}{2} \div 6 = \frac{1}{4}$ c per person, $\frac{1}{4} \times 10 = 2\frac{1}{2}$ c

Students may also use a combination of scaling down and scaling up:
$1\frac{1}{2}$ or $\frac{6}{4}$ c for 6 people
$\frac{1}{4}$ cup for 1 person
$\frac{10}{4}$ or $2\frac{1}{2}$ c for 10 people

3. 6 times greater; multiplying by 2 and then by 3 is the same as multiplying by 6:
smallest key $\times 2 \times 3$ = smallest key $\times 6$

Help students understand that both the enlargement and the reduction of the figure will be similar to the original and will also be similar to one another.

4. The result will be 75% of the original;
$0.5 \times 1.5 = 0.75$

Help students see that the scale factor for the enlargement is 1.5.

5. either 4 in. or $6\frac{1}{4}$ in.; $\frac{5}{10} \times 8 = 4$ in., $\frac{5}{8} \times 10 = 6\frac{1}{4}$ in.

The solution depends on whether the 5-in. side was the longer or shorter side of the original photo.

6. 10 times larger; lowest setting = $10 \times 4 = 40$, highest setting = $10 \times 40 = 400$, $400 \div 40 = 10$

7. smaller to larger: $f = \frac{32}{20} = 1.6$;
larger to smaller: $f = \frac{20}{32} = 0.625$

☑ **Informal Assessment** Ask: How could you test whether the two screens are similar? (Corresponding sides will form equal ratios.)

© Creative Publications 10874

3 ENLARGEMENTS & REDUCTIONS

Try It Out

1. Christine is using a copying machine to make a 75% reduction of a drawing of a car. How many inches shorter will the car be in the copy than in the original if the original car is 9 in. long?

2. Tom and Sonia use $1\frac{1}{2}$ c of rice for a dish that serves 6 people. How many cups of rice should they use for 10 people?

3. How many times greater would an enlargement of this key by a factor of 3 be than a reduction by a factor of $\frac{1}{2}$?

4. Stacey will use a copying machine to enlarge a drawing by 50%. Then he will reduce the enlargement by 50%. How will the result compare to the original?

5. Abdul has enlarged a photo to 8 in. by 10 in. One side of the original photo was 5 in. long. What was the length of the other side?

6. Helena is using a microscope that has one 10-magnification lens and a second lens with three magnification choices: 4, 10, or 40 times. How much larger does an ameba appear under the greatest magnification than under the least?

7. What factor relates the size of the two television screens?

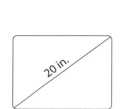

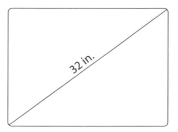

3 | ENLARGEMENTS & REDUCTIONS
Stretch Your Thinking

1. Find Out To help students understand the problem, ask:

- What question does the problem ask you to answer? (What copier settings can be used to reduce a drawing 10 in. long to 6 in. long?)

- What reduction factor does the customer need? (60% or 0.6)

- Can José use the copier just once to accomplish his goal? Explain. (No; the maximum he could accomplish with one use would be to create a 70% reduction.)

- If you make a reduction of a reduction, will the order of the reductions matter? (no)

- Would it make sense for José to consider enlarging the drawing first? (No; if you cannot use one reduction to get to 60%, you cannot get an even larger copy down to 60%.)

Solution Path
Use the fact that $8 \times 0.75 = 0.6$. First, reduce the original to 80%, making a copy that is 8 in. long. (10 in. $\times$ 0.8 = 8 in.) Then, reduce the copy to 75%, making another copy that is 6 in. long. (8 in. $\times$ 0.75 = 6 in.)

This is a good problem for the *make a table* strategy. Point out that the students are looking for two numbers within the given range of 70% to 120% that multiply to 60%. Students may find it easier to work with decimal equivalents.

2. The result is 84% of the original regardless of the order; $0.7 \times 1.2 = 1.2 \times 0.7 = 0.84$

☑ **Informal Assessment** Ask: Would the effect of José's experiment be the same regardless of the size of the original? (Students should see that the same result would occur regardless of the original's size.)

3. 9 in. by 12 in.; use the scale factor, $42 \div 28 = 1.5$, $8 \times 1.5 = 12$ in. long and $6 \times 1.5 = 9$ in. wide

4. 95 in.; use the scale factor, $15 \div 6 = 2.5$, $38 \times 2.5 = 95$ in.

Some students might choose to use proportions to solve this problem. This method, while acceptable, is not as efficient because students need to multiply and divide larger numbers.

5. $1\frac{3}{4}$ c; the flour is increased by a factor of $2\frac{1}{2}$, so increase the milk by the same factor to $6\frac{1}{4}$ c. There are 8 c in $\frac{1}{2}$ gal, so $1\frac{3}{4}$ c are left.

Some students may use the strategy *make it simpler* to replace the mixed numbers with whole numbers to find a solution path.

3 | ENLARGEMENTS & REDUCTIONS

Stretch Your Thinking

1. José works for the Copy Shop after school. Today, he has to reduce a drawing that is 10 in. long to exactly 6 in. The scale factors for the copy machine go from 70% to 120% in steps of 5%. How can José complete the job by making only two reductions?

2. José wanted to see what would happen if he made the maximum reduction on the copy machine followed by the maximum enlargement. What did he discover? Would the same be true if he reversed the order?

3. Nina has a photograph that measures 6 in. by 8 in. The inside of the frame she has is similar to the photograph and has a perimeter of 42 in. To what size must Nina enlarge her photograph to fit the frame?

4. If a banner similar to the one shown is made so that the shortest side is 15 in. long, what will the perimeter of the banner be?

16 in.

6 in.

DAVIS GYMNASTIC CLUB

16 in.

5. A chef has a cake recipe that uses $3\frac{1}{2}$ c flour and $2\frac{1}{2}$ c milk. She increases the recipe so that $8\frac{3}{4}$ c of flour are needed. If she has $\frac{1}{2}$ gal of milk, how much milk will be left?

© Creative Publications 10874

Problem Solving with Ratio and Proportion **33**

3 | ENLARGEMENTS & REDUCTIONS
Challenge Your Mind

1. **Make a Plan** To help students devise a plan for solving the problem, ask:

- To be sure you are getting the correct answer for part **a**, what do you need to know about the original garden? (its width)

- What strategy might you use to find the dimensions of the new garden? (Possible answer: *make a picture or diagram* strategy)

- How can you find the scale factor for part **b**? (Possible answer: divide the new width by the original width)

Solution Paths

a. The new garden will be 90 ft long and 135 ft wide. Students can use the *guess and check* strategy, keeping in mind that the new length and width must create a rectangle similar to the first.

b. The scale factor is 1.5;

$$\frac{\text{new width}}{\text{original width}} = \frac{90}{60} = 1.5,$$

$$\frac{\text{new length}}{\text{original length}} = \frac{135}{90} = 1.5$$

This is a good opportunity to show students an application of simple algebra.

Let f = scale factor

$$60 \times \text{width} = 5,400$$
$$(f \times 60) \times (f \times \text{width}) = 12,150$$
$$\frac{(f \times 60) \times (f \times \text{width})}{60 \times \text{width}} = \frac{12,150}{5,400}$$
$$f^2 = 2.25$$
$$f = 1.5$$

c. The area has increased by $\frac{12,150}{5,400} = 2.25$. Discuss the relationship between the increase in length or width (1.5 times) and the increase in area (2.25 times). Students may recognize that 2.25 is the square of 1.5.

---

2. **a.** Each homesite will be a rectangle 100 ft by 240 ft, with an area of $100 \times 240 = 24,000$ ft^2.

Students can find the dimensions by using a proportion or by division: $1,000 \div 10 = 100$ ft; $2,400 \div 10 = 240$ ft

b. The ratio is 1:100; compare the area of a homesite, 24,000 ft^2, to the area of the entire land, $1,000 \times 2,400 = 2,400,000$ ft^2.

c. There will be 100 homesites; $2,400,000 \div 24,000$.

✔ **Informal Assessment** Ask: How are all of the rectangular homesites related to each other? How are they related to the entire property? (The homesites will all be congruent rectangles, and they will all be similar to the entire piece of land.)

3 ENLARGEMENTS & REDUCTIONS

Challenge Your Mind

1. A rectangular garden was 5,400 ft² and 60 ft long. Now the garden is 12,150 ft², and the new rectangle is similar in shape to the old one.

a. What are the dimensions of the new garden?

b. By what factor have the length and width of the original garden grown?

c. By what factor has the area increased?

2. The city of Soundview is selling homesites from a piece of land that the city owns. The land for homes is a rectangle 1,000 ft by 2,400 ft. Land for streets and sidewalks will be cut from people's properties after they have been bought.

a. The ratio of each homesite's dimensions to the corresponding dimensions of the entire piece of land is 1:10. What will the area of each homesite be?

b. What will be the ratio of the area of a homesite to the area of the entire piece of land?

c. How many homesites will there be?

© Creative Publications 10874   Problem Solving with Ratio and Proportion   **35**

3 | ENLARGEMENTS & REDUCTIONS
Wrap It Up

The Art of Enlargements

Discuss the figure with the class. Ask:

- What appears to be the relationship between triangle *ABC* and triangle *A'B'C'*? (They appear to be similar.)

- What does the square symbol in the corner of the triangles mean? (The symbol represents a right, or 90°, angle.)

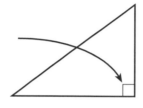

Solutions

a.
PA = 24 mm	*PB* = 12 mm	*PC* = 24 mm
PA' = 72 mm	*PB'* = 36 mm	*PC'* = 72 mm
AB = 18 mm	*AC* = 24 mm	*BC* = 30 mm
A'B' = 54 mm	*A'C'* = 72 mm	*B'C'* = 90 mm

b. a factor of 3

c. perimeter of triangle *ABC* = 72 mm, perimeter of triangle *A'B'C'* = 216 mm

d. The relationship of the perimeters is the same as the ratio between corresponding sides.

e. The area of triangle *A'B'C'* is exactly 9 times the area of triangle *ABC*.

Possible Solution Paths

b. Divide the length of a projection segment (for example *PA'*) by the length of the original segment (*PA*) to find the factor, 3.

c. Add the lengths of the sides of triangle *ABC* and *A'B'C'*. Or find the perimeter of triangle *ABC* and multiply it by 3 to get the perimeter of triangle *A'B'C'*.

d. Divide the perimeter of the larger triangle by that of the smaller triangle.

e. Students can use a tracing of triangle *ABC* to determine that they would need about 9 copies to cover triangle *A'B'C'*. Some students may realize that the areas are related by the square of the scale factor.

Assessment Rubric

3 The student correctly measures the lengths of the segments, choosing an appropriate unit; determines the enlargement factor, accurately finds the perimeter; and gives a reasonable estimate for the relationship of the areas.

2 The student correctly measures the lengths of the segments, choosing an appropriate unit; determines the enlargement factor; finds the perimeters and shows some understanding of the relationship of perimeters to corresponding sides; and makes some estimate for the relationship of the areas.

1 The student has some success measuring the length of the segments; gives a reasonable answer for the enlargement factor; can find the perimeters with some assistance but has little success relating perimeters to corresponding sides; and cannot make a reasonable estimate for the relationship of the areas.

0 The student can measure the length of the segments but does not show any understanding of the concept of scale factors or enlargements.

---

3 | ENLARGEMENTS & REDUCTIONS

Wrap It Up

The Art of Enlargements

When artists paint pictures of three-dimensional objects, they need a way to make a flat surface appear to have depth. One way to do this is to use projections.

The figure at the right shows a projection of triangle *ABC* from point *P*.

Triangle *A'B'C'* is an enlargement of triangle *ABC*.

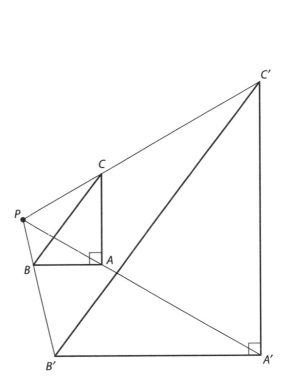

a. Measure the lengths of the following segments in millimeters:

PA	*PA'*	*AB*	*A'B'*
PB	*PB'*	*AC*	*A'C'*
PC	*PC'*	*BC*	*B'C'*

b. By what factor has triangle *ABC* been enlarged?

c. Find the perimeter of each triangle.

d. How does the relationship of the perimeters compare to the relationship of any pair of corresponding sides?

e. What is the relationship between the areas of the two triangles?

---

4 | SCALE DRAWINGS
Introduction

In this section, students solve problems relating to drawings that are scaled up or down. All scale drawings are examples of similar figures. If an object is large, a scale drawing of it is usually a reduction. A drawing of a small object, such as a computer chip, that has been scaled up is an example of an enlargement.

Understanding Scale Drawing Problems

Students should be familiar with the concepts of similar figures, including enlargements and reductions. They need to understand that enlarging or reducing a figure changes all lengths by the same factor but changes the area by a different factor.

Example:

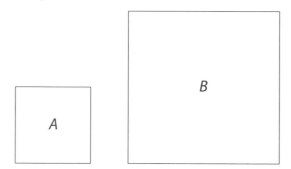

The scale factor of a side of square *B* to a side of square *A* is 2:1. However, the area of square *B* is four times the size of the area of square *A*.

Encourage students to get into the habit of thinking critically when they encounter expressions such as "twice as big" or "half as large." What aspect of the object is twice as big or half as large—the length? The area? The volume?

As students prepare to solve scale drawing problems, they should ask themselves:

- What question am I trying to answer?

- What does each part of the scale factor represent?

- How are the two parts of the scale factor related?

- How are the size of the scale drawing and the size of the actual figure related?

- How could I use the scale factor to solve the problem?

  People often show scales in the following form: 1 cm = 5 m. You may wish to point out to your students that, technically, it is incorrect since 1 cm does not equal 5 m. Throughout this book, scales are shown as ratios, such as 1 cm:5 m.

Solving Scale Drawing Problems

Students will frequently make use of proportions to solve scale drawing problems. Help your students focus on each ratio in the proportion when setting up a proportion. If they understand the relationships between the numbers, solving the proportion will follow easily. They should be careful when identifying the relationship between the scale and the measurements in a scale drawing.

Students will encounter two important considerations when working with scale drawings.

- Is the scale drawing an enlargement or a reduction?

  Example: Suppose a scale drawing of an ant uses the scale 5:1. Since ants are generally small, it is reasonable to assume that the number 5 refers to the drawing and not the ant. Help students focus on building proportions that make sense.

- Does the problem ask for the size of the figure (either the original or scale drawing) or for the scale?

  Example: Consider an 18-ft-long car and a scale drawing that is 4 in. long. In the scale drawing, what does 1 in. represent? The question asks students to describe the scale (1 in. represents $4\frac{1}{2}$ ft). Discuss how the scale relates the size of the original figure to the size of the scaled figure.

Assessment

☑ **Informal Assessment** A suggestion for informal assessment will be found on each Try It Out, Stretch Your Thinking, and Challenge Your Mind page. The recommended question will help focus students' attention on one part of the problem-solving process.

Assessment Rubric An assessment rubric is provided for each Wrap It Up. Students' completed work may be added to their math portfolios.

Thinking About Scale Drawings

These problems help students use their number sense and estimation skills as preparation for solving scale drawing problems. Present one problem a day as a warm-up. You may choose to read the daily problem aloud, write it on the board, or create a transparency.

1. **A scale drawing of a hexagon is drawn so that the scale ratio is 3:1. How many sides will the scale drawing have?**
   (6 sides; scale drawings do not change the number of sides or angles.)

2. **You want to make a scale drawing of a 170-in.-long car. If your scale is 1 inch:10 inches, will it fit on a piece of notebook paper?**
   (No; the drawing will be 17 in. long. Most notebook paper is only 11 in. long.)

3. **On graph paper, draw two squares: one whose side is 2 units long and one whose side is 4 units long. How do the lengths and areas of the two squares compare?**
   (The side length of the larger square is twice that of the smaller square; the area is four times greater.)

4. **Some people make clothing by using a pattern to cut pieces of fabric. What, do you think, is the scale of the pattern?**
   (1:1, since the clothing will be the size of the pattern. Scale drawings can sometimes be the same size as the originals.)

5. **The Statue of Liberty is 151 ft tall from her feet to the top of her torch. What is a reasonable guess for how many times more than life-size the statue is?**
   (Possible answer: about 25 times life-size)

4 | SCALE DRAWINGS

Solving Problems with Scale Drawings

This lesson allows students to apply the concept of scale drawings to measurement. The solution requires students to use what they know about the relationship of two linear measures: feet and yards.

Using the Four-Step Method

Find Out

○ Students must determine whether the salesperson told the Shahs the correct amount of carpeting needed and, if not, find the correct amount.

● Students need to know the length and width of the room to calculate its area. It may be helpful to review the difference between area and perimeter.

○ The room dimensions are given in feet but carpet is sold by the square yard.

● The scale 1 inch:6 ft means that every 1 in. on the scale drawing corresponds to 6 ft in the actual room.

○ Measure the length and width in the drawing. Then use proportions or scaling up to find the length and width of the actual room.

● Students can estimate that the area of the room is about 6×4, about 24 yd^2.

Make a Plan

○ Find the area of the room and compare it to 20 yd^2.

Solve It

● The Shahs' room requires 22 yd^2 of carpeting. So, the salesperson was incorrect.

Two possible solution paths are given.

	Scale Drawing	Actual
Scaling Up	1 in.	6 ft
	2 in.	12 ft
	$2\frac{3}{4}$ in.	$16\frac{1}{2}$ ft
	$12 \text{ ft} \times 16\frac{1}{2} \text{ ft} = 198 \text{ ft}^2$	
	$198 \text{ ft}^2 \div 9 = 22 \text{ yd}^2$	

Solving Proportions	$\frac{1}{6} = \frac{2.75}{x} \qquad \frac{1}{6} = \frac{2}{y}$
	$x = 16.5 \text{ ft} \qquad y = 12 \text{ ft}$
	$16.5 \div 3 = 5.5 \text{ yd and } 12 \div 3 = 4 \text{ yd}$
	$5.5 \times 4 = 22 \text{ yd}^2$

Look Back

○ Have students compare their solutions to their estimates to see if these are in reasonable agreement. Ask whether they think estimating helped them solve the problem.

● A way to check the solution is to solve by a different method. Encourage students to share their solution methods in small groups.

4 SCALE DRAWINGS

Solving Problems with Scale Drawings

Mr. and Mrs. Shah made this scale drawing of a room they want to carpet. A salesperson told them that they will need 20 yd² of carpeting. Do you think the salesperson is correct? If not, how much carpeting do you think the Shahs will need?

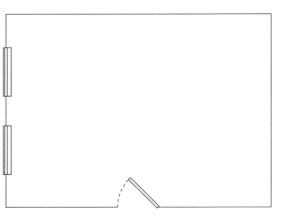

Scale: 1 inch = 6 feet

Find Out	
	○ What question must you answer to solve the problem?
	• In order to carpet a room, what room measurements are you interested in?
	○ Why do you need to know the relationship between feet and yards to solve the problem?
	• What does the scale 1 inch: 6 ft mean?
	○ How could you use the scale drawing to find the measures (length and width) of the Shahs' room?
	• Estimate the area of the room in square yards.
Make a Plan	○ How can you use the length and width of the room to check what the salesperson told the Shahs?
Solve It	• Complete the solution, using the plan that you have made. Keep a record of your work.
Look Back	○ How does your solution compare to your estimate?
	• How would you check to see whether your solution is correct?

4 | SCALE DRAWINGS
Try It Out

1. **Find Out** To help students understand the problem, ask:

- What does the problem ask you to find? (the length of the actual space shuttle)

- How could knowing the scale of the drawing help you solve the problem? (You can measure the length in the drawing and use the scale to calculate the actual length.)

- What is the relationship of the numbers in the scale? (For every 0.5 in. of the drawing, the actual space shuttle is 20 feet long.)

Solution Path

The space shuttle is approximately 120 ft long;

$$\frac{0.5 \text{ in.}}{20 \text{ ft}} = \frac{3 \text{ in.}}{x \text{ ft}} \text{ or } \frac{0.5 \text{ in.}}{3 \text{ in.}} = \frac{20 \text{ ft}}{x \text{ ft}}, \; x = 120$$

2. Approximately 40 ft greater; the drawing is about 1 in. longer than it is wide, so the shuttle is about 2×20 ft = 40 ft longer than it is wide.

✓ **Informal Assessment** Ask: What is a good estimate of the ratio of the width of the shuttle at its widest part to its length? (The ratio of width to length is approximately $\frac{80}{120}$, or $\frac{2}{3}$.)

3. The shuttle is about $\frac{1}{3}$ as long as the building is tall.

	Scale Drawing	Actual
One way to solve this problem is to scale up.	0.25 in.	20 ft
	0.50 in.	40 ft
	0.75 in.	60 ft
	1 in.	80 ft
	4 in.	320 ft

The building is 320 + 60 = 380 ft tall. It is about 380 − 120 = 260 ft taller than the space shuttle is long. $\frac{120}{380}$ is about $\frac{1}{3}$.

4. 6 in. greater; subtract $1\frac{3}{4} - 1 = \frac{3}{4}$ in., then multiply $\frac{3}{4} \times 8 = 6$ in.; or multiply $1\frac{3}{4} \times 8 = 14$ in. and $1 \times 8 = 8$ in., then subtract 14 − 8 = 6 in.

5. About 60 calories; the ratio of the areas of the slices equals the ratio of the calories for the slices, so write and solve a proportion.

$$\frac{\frac{1}{8} \times \pi \times 7^2}{\frac{1}{8} \times \pi \times 4^2} = \frac{183}{x}, \; x \approx 60 \text{ cal}$$

Students may realize that the ratio of the areas is equal to the ratio of the squares of the radii. Discuss why this is true. Explore any other solution strategies suggested by students. Answers may vary depending on what value students use for π or if they use the π key on the calculator.

6. 12 in., 18 in., and 21 in.; multiply 12 in. × 1 = 12 in., 12 in. × $1\frac{1}{2}$ = 18 in., 12 in. × $1\frac{3}{4}$ = 21 in.

4 | SCALE DRAWINGS

Try It Out

1. This is a scale drawing of the space shuttle *Discovery*, the first reusable spacecraft. What is the approximate length of the actual space shuttle?

2. Approximately how many feet greater is the length of the actual shuttle than its width at its widest part?

3. A scale drawing of the Longfellow Towers in Boston, Massachusetts, uses the scale 0.25 inch:20 ft. The height of the building in the drawing is 4.75 in. How does the actual length of the space shuttle compare to the actual height of the building?

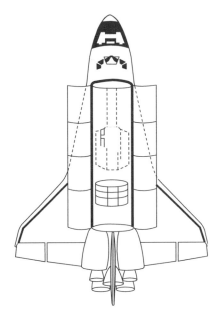

Scale: 0.5 inches = 20 feet

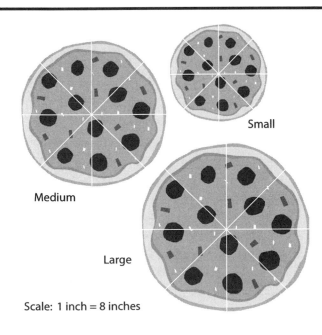

Small

Medium

Large

Scale: 1 inch = 8 inches

4. Pete's Pasta and Pizza sells pizza pies in three sizes. These are scale drawings for each pie. How much greater is the diameter of Pete's large pie than the diameter of Pete's small pie?

5. Pete cuts all his pizzas into eighths. One slice of the large cheese pizza has 183 calories. About how many calories does a slice of the small cheese pizza have?

6. Suppose Pete changed the scale to 1 inch:1 ft but the drawings remained unchanged. What would be the sizes of Pete's actual pizzas?

4 SCALE DRAWINGS

Stretch Your Thinking

1. **Find Out** To help students understand the problem, ask:

- What does the problem ask you to do? (Make a scale drawing and find out how high above the ground the actual kite is.)

- What information are you given? (The scale of the sketch is 1 cm : 40 m; in the sketch, the kite is 3 cm from the ground, and the length of the string is 5 cm.)

- Do you need the sketch to find the actual height of the kite? Explain.
  (No, you can use the scale to find the actual height: 1 cm : 40 m = drawing cm : actual m.)

Solution Paths

a. Sketches may vary in the orientation of the kite. A sample is shown. Measurements of the kite height and string length should be accurate.

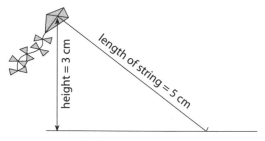

Scale: 1 cm = 40 m

b. 120 m
   Students can scale up: 1 cm : 40 m so 3 cm : 120 m or students can solve a proportion:

   $$\frac{1}{40} = \frac{3}{x}, \ x = 120 \text{ m}$$

---

2. a. About 49 in.; use a proportion,

   $$\frac{1}{20} = \frac{x}{986}$$
   $$x = 49.3$$

   b. About 73 in.; 1,454 ÷ 20 = 72.7

3. a. About 18.5 in.; use a proportion,

   $$\frac{1}{2.7} = \frac{x}{50} = 18.5$$

   b. 1 inch : 1.35 ft; 1 in. should correspond to half of 2.7 ft.

   ☑ **Informal Assessment** Ask: In the movie, King Kong appeared to be standing on the Empire State Building. The actual building is 1,250 ft tall. If a model of the Empire State Building were built to the same scale as King Kong, about how tall would the model be? (About 39 ft. The actual building is 25 times taller than Kong's 50-ft height. So, 25 × 18.5 in. = 462.5 in. ÷ 12 = 38.54 ft.)

---

4. 16 beanbag animals; 5 × 0.3 m = 1.5 m, or 150 cm, 150 ÷ 9 = 16.67

Students will need to know that there are 100 cm in 1 m to solve the problem.

5. 3 cm; the queen honeybee is 1 cm × $\frac{6}{4}$ = 1$\frac{1}{2}$ cm, the queen wasp is 2 × 1$\frac{1}{2}$ cm = 3 cm

4 | SCALE DRAWINGS

Stretch Your Thinking

1. A scale drawing shows a kite attached to a string that is fastened to the ground with a stake. The height of the kite above the ground is 3 cm. The length of the string is 5 cm.

a. Make a sketch of the drawing using the scale 1 cm:40 m.

b. How high above the ground was the actual kite?

---

2. The scale for a model of the Eiffel Tower in Paris, France, is 1 inch:20 ft. The Eiffel Tower is actually 986 ft tall.

a. How tall is the scale model?

b. If a model of the 1,454-ft-high Sears Tower in Chicago, Illinois were built to the same scale, how tall would the model be?

3. In the original version of the movie *King Kong* made in 1933, King Kong appeared to be 50 ft tall. King Kong was actually a model that was built to a scale of 1 inch:2.7 ft.

a. Approximately how tall was the model?

b. What scale would the movie maker have used to make the model about twice as large?

---

4. In Scott's scale drawing of his shelf unit, the scale is 1 cm:0.3 m. Scott wants to place his beanbag animal collection on the shelf. The average width of each animal is 9 cm. What is the greatest number of beanbag animals that will fit across the shelf?

Scale: 1 cm = 0.3 m

5. The queen honeybee below has been drawn to a scale of 4 cm:1 cm. If a queen wasp is twice as long as a queen honeybee, what is the actual length of a queen wasp?

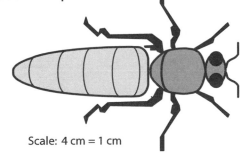

Scale: 4 cm = 1 cm

4 | SCALE DRAWINGS
Challenge Your Mind

1. **Find Out** To help students understand the problem, ask:

- What does the problem ask you to do? (Find a reasonable scale to use to make a scale drawing on $8\frac{1}{2}$-by-11 inch paper and create a scale drawing using that scale.)

   Make a Plan To help students decide on a strategy to follow, ask:

- How will you pick a reasonable scale? (Compare how big you want the drawing to be, the size of the paper, and the actual size of the whale.)

- What must you do after you decide on a scale to make the scale drawing? (Use the scale to find the dimensions of the whale in the drawing.)

- How will you decide how wide to make the whale in your drawing? (Possible answer: Use the scale

to determine the dimensions of the whale to be drawn, make a rectangle with those dimensions, then sketch the whale.)

Suggest that students use grid paper to make their scale drawings.

Solution Paths

a. Scales may vary. Since the longest blue whale ever recorded was 110 ft and notebook paper is 11 in. long, students' scales should be near 1 inch:11 ft. (If the scale were 1 inch:10 ft the drawing would extend to the end of the paper and would be difficult to draw.)

b. The scale drawing will depend on the scale chosen in part **a**. Students should provide their calculations along with their drawings.

---

2. a. 16 m long; 8 cm × 200 = 1,600 cm; 1,600 cm ÷ 100 = 16 m

   Some students may mentally convert 8 cm to 0.08 m and multiply 0.08 by 200 to find the length in meters.

   b. 3.2 m wide;
   1.6 cm (the width of the scale drawing) × 200 = 320 cm and 320 ÷ 100 = 3.2 m

   Some students may use the fact that the length is 5 times the width: $x = 16 \div 5$, $x = 3.2$ m.

c. Students can choose a scale factor that will result in an enlargement of the given scale drawings. Have students share their drawings and justify their scales by writing proportions.

   ✓ **Informal Assessment** Ask: If you want to make a scale drawing that is half as large as Rafael's, what scale would you use? Why? (1:400; when the scale is 1:200, 1 cm on the actual object equals 0.005 cm on the drawing; for a drawing half as large, 1 cm on the actual object must equal $\frac{1}{2}$ of 0.005 cm, or 0.0025 cm, on the drawing)

4 | SCALE DRAWINGS

Challenge Your Mind

1. The length of the longest blue whale ever recorded was 110 ft.

a. What is a reasonable scale to use if you want to make the largest possible scale drawing of a blue whale that will fit on a sheet of notebook paper?

b. On notebook paper, make a scale drawing of the longest blue whale ever recorded. Use your scale.

2. Rafael builds model railroad cars based on actual cars. He made the scale drawings below from measurements he took of an aluminum-sided hopper. Aluminum cars weigh less than steel cars and can carry up to 50 percent more. Rafael's scale is the ratio 1:200.

a. How many meters long is the actual railroad car that Rafael used as his model?

b. The length of the car is 5 times the width. How many meters wide is the car?

c. Make an enlargement of Rafael's scale drawings. What scale will you use?

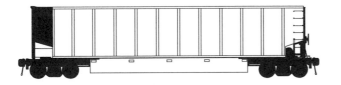

4 SCALE DRAWINGS

Wrap It Up

Play Ball!

Part 1 Draw a sketch of a Little League field on the board.

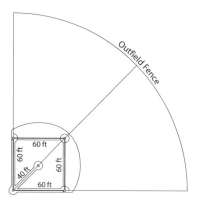

Ask:

- Where would you place your drawing of the diamond to make the largest possible drawing? (Holding the paper vertically, at the bottom left of the paper. This will allow the greatest distance for the outfield corner to corner measures.)

- How will you decide on a scale? (Possible answer: Draw the 203-ft distance to the outfield fence along the $8\frac{1}{2}$-in. length. The scale would derive from $8\frac{1}{2}$ inches:203 ft. That could translate into a convenient scale of 1 inch:30 ft.

- Whatever scale you choose, approximately how many times greater will the line from home plate to the outfield fence be than the line from home plate to the pitcher? (approximately 5 times greater)

Solution

Refer to the diagram above. For a scale of 1 inch:30 ft, the distances between bases would be 2 in., the distance from the pitcher's mound to home plate would be $1\frac{1}{3}$ in., and the distance from home plate to the outfield fence would be approximately $6\frac{3}{4}$ in.

Part 2 Have students work with partners to read, critique, and revise their problems. Present the most challenging problems to the class and solve them as a whole-class activity.

If students get stuck, you might prompt them with questions such as:

- If a fielder catches a ball at third base and throws to first base, how long is the throw?

- What is the distance traveled from home plate to second base by a batted ball?

- What is the distance between players on first base and third base?

Assessment Rubric

3 The student determines a reasonable scale, accurately makes a scale drawing of the baseball field including the scale, writes and answers two problems that require use of the scale drawing and scale to solve.

2 The student determines a reasonable scale, makes a reasonably accurate scale drawing of the baseball field including the scale, writes and answers at least one problem about the field or the scale drawing regardless of whether the scale drawing and actual field must be used to solve.

1 The student chooses a scale, but draws a picture of the baseball field rather than making a scale drawing, writes a problem related to the drawing or actual field but not connecting the two.

0 The student cannot choose a reasonable scale, draws a picture that does not reflect the distance relationships given, may formulate a question involving the drawing or actual field that does not require either to solve.

Name _____

4 | SCALE DRAWINGS
Wrap It Up

Play Ball!

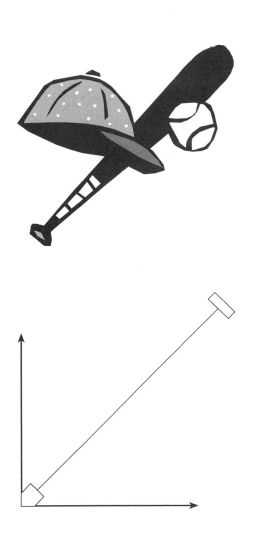

Each year the Little League World Series is played in Williamsport, Pennsylvania. The bases are 60 ft apart. The pitcher's mound is 40 ft from home plate. The outfield fences are all 203 ft from home plate.

Part 1

Decide on a reasonable scale for drawing the rest of the field. Then copy the diagram of home plate and the start of the field. Draw the field and write the scale you have chosen.

Part 2

Make up two problems that need your scale drawing for their answers. Prepare solutions to your questions. Challenge a partner or other classmate to solve the problems.

5 | MAPS
Introduction

In this section, students solve problems relating to maps. Today, we assume that for any trip we are planning, whether around the corner or around the world, there will be an accurate map available. Most formal maps are drawn to a scale. However, most maps are not scale drawings because they are two-dimensional representations of a curved three-dimensional world. Maps people draw for driving directions, seating plans, and transit system maps will not usually be drawn to a scale.

Understanding Map Problems

Success at reading and understanding maps requires familiarity with rates and scale drawings. Since the scale of a map usually uses two different units, the scale is generally a rate. When maps are drawn to a scale, they have the characteristics of a scale drawing. In general, the scale factor used to create the reduction can be determined from the scale.

As students consider map problems, the questions they should be asking themselves include:

- What question am I trying to answer?
- Are the scale parts written in the same units? If not, how are the units related?
- How can I understand the scale as a ratio?
- What does each part of the ratio represent?
- How can I use the scale to solve problems about distance on the map?
- How can I create a scale from information in the map?
- How can I use the information in the problem to check whether my solution is reasonable?

Example:

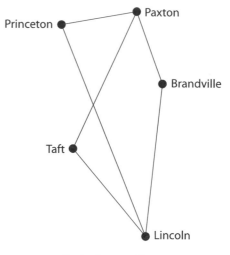

Scale: 1 cm = 1 km

Name two locations on the map and have students tell the distance between the two points. Encourage students to write down the distance they measure with a unit. Then have them use this and the map scale to write a proportion to solve for the actual distance.

Solving Map Problems

Solving problems involving maps requires an understanding of enlargements, reductions, and scale drawings.

In general, map problems ask students either to use a scale to find distance on the map or to determine the scale from the map. Students need to recognize the type of problem that they are working with and the kinds of information that they will need to solve the problem.

- In finding distance from a map, students need to interpret and use the map scale. This is especially important for setting up proportions correctly.

- To find the map scale if none is given, students will need to know at least one actual distance. They will need to decide on an appropriate unit for measuring distances between points on the map that correspond to known actual distances. With that information, they can create a scale.

Assessment

✓ **Informal Assessment** A suggestion for informal assessment will be found on each Try It Out, Stretch Your Thinking, and Challenge Your Mind page. The recommended question will help focus students' attention on one part of the problem-solving process.

Assessment Rubric An assessment rubric is provided for each Wrap It Up. Students' completed work may be added to their math portfolios.

Thinking About Maps

These problems help students use their number sense and estimation skills as preparation for solving map problems. Present one problem a day as a warm-up. You may choose to read the daily problem aloud, write it on the board, or create a transparency.

1. **You want to draw a map of California on an $8\frac{1}{2}$-by-11-in. sheet of paper. You want the map to cover as much of the paper as possible, using 1 inch as part of the scale. What do you need to consider?**
(the state's actual length and width; the ratio between the state's actual length and the length of the paper; the state's maximum width compared to the paper's width once a scale has been chosen based on the length)

2. **Which scale is on the larger map of the United States: 1 inch:50 mi or 1 inch:100 mi? Explain how you decided.**
(1 inch:50 mi; possible explanation: A lesser distance, 50 mi, is represented by 1 in., so it is on the larger map.)

3. **A map of Texas drawn to a scale of 1 inch:50 mi is placed on an overhead projector and enlarged on a screen. What part of the scale in the enlargement is still accurate? What must be changed?**
(The actual distance between points is correct; the units of the scale must be changed.)

4. **A company prints two maps of Canada. One map has a scale of 1 cm:500 km. The other has a scale of 5 cm:1,000 km. Which map is smaller? Explain how you know.**
(The scale in the second map is equivalent to 1 cm:200 km, so the first map is smaller.)

5. **You and a friend reduce a map to 75% of its original size. The scale for the map was 1 inch:400 mi. Your friend says that in the reduction, 1 in. represents less than 400 mi. Is that correct? Why do you think so?**
(No, as the map gets smaller, 1 in. represents a greater distance.)

5 | MAPS

Solving Problems with Maps

In this section, students will use their understanding of ratio, proportion, and scale drawings to solve a problem involving a map.

Using the Four-Step Method

Find Out	○ Find a scale that would be reasonable using the only given distance (400 mi) and use that scale to locate Phoenix, Arizona on the map.
	● Facts to consider include the map distance between Los Angeles and San Francisco, the actual distance between the two cities, what unit is appropriate for the size of the given map, and how the unit chosen will relate to the distance between the two cities.
Make a Plan	○ Students must relate the actual distance of 400 mi to the distance between Los Angeles and San Francisco that they measure on the map.
	● Students can use a ruler to mark numerous points in Arizona that are 400 mi from Los Angeles and 700 mi from San Francisco and look for an intersection.
Solve It	○ an infinite number; an infinite number; at most two
	● a. 1 inch:400 mi
	b. Students will mark points that represent 400 mi from Los Angeles and points that represent 700 mi from San Francisco. As they continue to mark points, they should notice that the marks are forming circles and that the circles will meet. Some students will recognize that by drawing a circle with Los Angeles at the center and a radius of 1 in., all points on the circle will be 400 mi away. Likewise, all points on a circle with a $1\frac{3}{4}$-in. radius centered on San Francisco will be 700 mi away. Students will then have to choose the correct point of intersection, using the map.
Look Back	○ Students could work backward to check that their scale is correct. They can assume that it is correct, then use it to find the distance between San Francisco and Los Angeles.
	● Students should use their knowledge of American geography to decide which location is reasonable and which does not make sense. Phoenix belongs in Arizona, not in the Pacific Ocean. Students should check that Phoenix is 400 mi from Los Angeles and 700 mi from San Francisco.

5 | MAPS

Solving Problems with Maps

Elizabeth drew the map at the right, but she forgot to include a scale. San Francisco is just over 400 mi from Los Angeles.

a. Find a scale that would be reasonable for Elizabeth's map. Use an inch ruler.

b. Phoenix, Arizona is just under 400 mi from Los Angeles and just over 700 mi from San Francisco. Use your scale to locate Phoenix on the map.

Find Out	○ What must you find to solve the problem?
	• What facts must you consider as you think about making a scale for the map?
Make a Plan	○ How can you find a reasonable scale?
	• How could you use your scale to locate Phoenix on the map?
Solve It	○ How many places can be 400 mi from Los Angeles? How many points can be 700 mi from San Francisco? How many places can meet both of those conditions?
	• Complete the solution, using the plan that you have made. Keep a record of your work.
Look Back	○ How did you choose the location of Phoenix from the possible locations you found?
	• How can you check your solution?

5 | MAPS
Try It Out

1. Find Out To help students understand the problem, ask:

- What does the problem ask you to find? (which plan is less expensive)

- What does the scale of the map tell you? (1 in. on the map represents 100 mi)

- What information can you get from the map? (approximate distances between cities)

Solution Path

Plan B costs about $12.50 less.

Mileage: Cleveland to Cincinnati ≈ 250 mi

Plan A: $50 + $0.10 × (250 − 200) = $55.00

Plan B: $20 + $0.15 × (250 − 100) = $42.50

2. Both plans would cost about the same amount.

Round-trip mileage between Cleveland and Cincinnati is about 500 mi.

Plan A: $50 + $0.10 × (500 − 200) = $80.00

Plan B: $20 + $0.15 × (500 − 100) = $80.00

3. 3 hr more;

Flying: $1\frac{1}{2} + \frac{1}{2} + \frac{1}{2} + 1 = 3\frac{1}{2}$ hr;

Driving: $1\frac{5}{8} × 190 ≈ 309$ mi, $309 ÷ 50 = 6.2$ hr, or about 6 hr;

$6 − 3\frac{1}{2} = 2\frac{1}{2}$ hr

✔ **Informal Assessment** Ask: If you know the distance between Jacksonville and Miami, but do not know the scale, how could you find a scale for the map? (Measure the map distance using any convenient unit and find the distance represented by 1 unit: $\frac{\text{map distance}}{\text{actual distance}} = \frac{1 \text{ unit}}{\text{distance represented by 1 unit}}$.)

4. About $230;

2 × $139 = $278 round-trip fare,

618 ÷ 20 = 30.9 gal of gasoline needed,

30.9 × $1.42 = $43.88;

$278 − $44 = $234

You may wish to discuss whether there are hidden costs for driving or flying.

5. Between $2\frac{1}{2}$ and 3 hr; 142 ÷ 50 = 2.85 hr

6. About an extra $1\frac{1}{2}$ hr; Jacksonville to Orlando is about 120 mi and Orlando to Miami is about 200 mi, 200 − 120 = 80 mi, 80 ÷ 50 = 1.6 hr

5 | MAPS

Try It Out

1. Creative Car Rentals offers two plans. With plan A you pay $50 and then $0.10 per mile after 200 mi. With plan B you pay $20 and then $0.15 per mi after 100 mi. For a one-way rental from Cincinnati to Cleveland, which plan would cost less? How much less?

2. What if your plans change in Cleveland and you need to drive the car back to Cincinnati? Which plan would you choose? Explain.

Scale: 1 inch = 125 miles

---

3. Mr. and Mrs. Leoni can't decide whether to fly or drive from Jacksonville to Miami. If they fly, they will need $1\frac{1}{2}$ hr to get to the airport, $\frac{1}{2}$ hr to rent a car at the Miami airport, and $\frac{1}{2}$ hr to drive into Miami from the airport. The flying time is 1 hr. If they drive, they will be able to average 50 mi/hr. How much longer will it take them to drive than to fly?

4. The round-trip airfare from Jacksonville to Miami is $139 per person. The Leonis' car averages 20 mi/gal and gasoline costs about $1.42 per gal. What is the approximate difference in cost between flying and driving?

5. If they drive, Mr. and Mrs. Leoni will stop in Orlando. At 50 mi/hr, how long would it take to drive from Jacksonville to Orlando?

6. How much longer would it take to drive from Orlando to Miami than from Jacksonville to Orlando if you drive at 50 mi/hr?

Scale: 1 inch = 190 miles

5 MAPS

Stretch Your Thinking

Before students attempt to solve the problem, tell them that scales without units mean that both parts of the scale are in the same units. For example, a scale of 1 ft:50 ft can also be written as 1:50.

1. Find Out To help students understand the problem, ask:

- How can you use the distance between Cauley and Barton to create a scale for Keith's map? (Measure the map distance between the two towns. Then write a scale that relates the map distance, 1 in., to the distance between the towns, 10 mi. Use the map with the 1:200,000 scale to find the actual distance.)

Make a Plan To help students decide on a strategy to follow, ask:

- What can you do to compare the scale that you will write for Keith's map to the scale of the map Keith found? (Change the units of one of the scales so that both scales compare the same units.)

Solution Path

Change the units in the scale of Keith's map so that both units of the scale are the same.

The two towns are about 3×1 in., or 3 in. apart on the second map:
1 mi = 5,280 ft and 1 ft = 12 in.
$10 \times 5,280 \times 12 = 633,600$ in.
So, 1 inch:10 mi $\Rightarrow$ 1:633,600.

Compare the scales of the two maps:
A map with a scale of 1:633,600 is approximately one third the size of a map with a scale of 1:200,000. (On the smaller map, each unit represents over 600,000 units. On the larger map, a unit represents only 200,000 units.)

☑ **Informal Assessment** Ask: Which map do you think is larger? Why do you think so? (The second map is larger since on it, 1 unit covers only 200,000 like units. On Keith's map, 1 unit represents 633,600 units.)

2. He can use the scale to estimate the length and width of each state. Since each state is almost rectangular, he can use the area formula for a rectangle.

3. Colorado is larger than Wyoming; about 5,000 to 6,000 mi^2

The actual areas are 103,595 mi^2 (Colorado) and 96,989 mi^2 (Wyoming). Students' estimates may be greater than the actual for each state since the shape is not exactly rectangular.

4. Yes; Cassie's friend should expect her between 11:00 a.m. and noon; $100 \div 45 = 2.2$ hr.

5. Approximately 250 mi of the Colorado River run through Colorado.

Students may use a string to estimate nonlinear distances on the map. They can follow the path of the river with string, measure the length of string, and translate that into actual miles from the scale.

5 | MAPS
Stretch Your Thinking

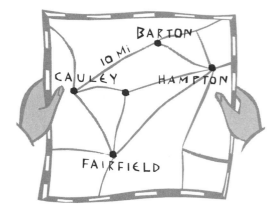

1. Keith made a map, drawn to scale, that included Cauley and Barton, two towns near his home. He forgot to put the scale on his map. Later he found the same two towns on a map with a scale of 1:200,000. About how far apart were the two towns on the map that Keith found?

2. Percy is writing a social studies report about the western United States and wants to check some of his data on a map. How could Percy use the map scale to estimate the areas of Colorado and Wyoming?

3. Use the map to estimate the areas of Colorado and Wyoming. Which is the larger state? Approximately how many more square miles is the area of the larger state than the smaller?

4. Cassie lives in Cheyenne and wants to visit her friend in Denver. If she leaves home at 9:00 a.m. and drives at an average of 45 mi/hr, can she be at her friend's house in Denver in time for lunch? At about what time should her friend expect her?

5. The Colorado River runs for a total of 1,450 mi from Rocky Mountain National Park in Colorado to the Gulf of California. About how many of those miles run through Colorado?

5 | MAPS

Challenge Your Mind

1. Find Out To help students understand the problem, ask:

- What questions does the problem ask you to answer? (About how far apart are the two cities; what is the average price you could pay for gasoline at 30 mi per gal; for what other cities might the claim be true?)

- To answer part **a**, why do you need to use the scale? (You need the scale to calculate the distance between the two cities.)

- How could knowing the scale of the drawing help you solve the problem? (You can measure the distance between the two cities, use the scale to calculate the actual distance, use the actual distance to calculate the number of gallons of gas, and then use that number to calculate the highest average price per gallon.)

Solution Paths

a. 250 mi; $\frac{1 \text{ in.}}{125 \text{ mi}} = \frac{2 \text{ in.}}{x \text{ mi}}$, $x = 250$ mi

b. The ad is only true if the highest average price you pay for gasoline is $1.20 per gal.

   $250 \div 30 \approx 8.3$ gal
   $\$10 \div 8.3 \approx \$1.20$ per gal

c. They could make this claim for other pair of cities, as long as the cities are 250 mi or less apart. For example: Pittsburgh, PA to Rochester, NY or Albany, NY to New York City, NY.

---

2. The distance from Pittsburgh, PA to Buffalo, NY is about 190 mi.

A scale of 1:8,000,000 means that for every 1 unit of map distance, there are 8,000,000 units of actual distance.

   $1.5 \times 8,000,000 = 12,000,000$ in.

There are 5,280 ft in 1 mi and 12 in. in 1 ft.

   $5,280 \times 12 = 63,360$ in. in 1 mi
   $12,000,000 \div 63,360 = 189.39$ mi

✓ **Informal Assessment** Ask: How do you interpret a scale when no units are given? (Both parts of the scale are in the same units, regardless of what the units are.)

5 | MAPS

Challenge Your Mind

1. A car company in Pennsylvania advertises that you can drive from Pittsburgh to Philadelphia in one of their cars for less than $10 of gasoline. They claim their cars get 30 mi/gal.

a. Estimate the distance between the cities using the scale on the map.

b. What is the average price you could pay for gasoline if the advertising claim is true?

c. Could the car company make this claim for any other pairs of cities? Explain why or why not.

NEW YORK

Rochester
Buffalo
Albany

Scranton

PENNSYLVANIA

Pittsburgh

Philadelphia

New York City

Scale: 1 inch = 125 miles

---

2. On an earlier printing of the map shown above, the scale was given as 1:8,000,000. Use that scale to estimate the distance between Pittsburgh and Buffalo. How does this estimate compare to the one above?

5 | MAPS
Wrap It Up

Vacation Time

As you discuss the Wrap It Up with the class, ask:

- How will you select a National Park? (Discuss the locations of some of the National Parks on the map. If a student selects a park not shown on the map, the student can mark its approximate location by consulting an atlas.)

- What is an itinerary? (a plan for the trip)

- How can you use the fact that the round trip in *part 1* must be between 500 and 1,000 mi to help you plan your itinerary? (The distance one way will be between 250 and 500 mi.)

- What restrictions must you consider as you plan your first trip? (Possible answer: maximum and minimum total distance of the trip, the time to be spent at each point of interest)

- What restrictions must you consider as you plan your second trip? (Students may have similar considerations to their first trip. This time, however, they must visit at least 5 different places and stay within 1,500 mi of the national park.)

Solutions

Part 1 Students may select one or two destinations along a single path since they may make only one round trip in the first problem. They will have to be certain that their round trip is between 500 mi and 1,000 mi. Encourage them to explain how they decided on their itineraries.

Part 2 Students will have to consider more distances and destinations in the second problem. They will have to be certain that their round trip is no more than 1,500 mi of the park they choose. This distance should allow a significant variety for each student's choice.

Possible Solution Paths

In both parts of the Wrap It Up students can measure each segment of the trip, use the scale to find the mileage, then add the results. They can also measure the entire length of the trip, then apply that length to the scale, using the scale just once.

Assessment Rubric

3 The student correctly plans both itineraries; shows the routes on the map; computes the mileage for each section of both trips, correctly showing how the mileage was computed for each itinerary.

2 The student plans both itineraries; shows the routes on the map; correctly computes most of the mileage for both trips, showing how the mileage was computed even if several computational errors occurred.

1 The student selects the correct number of destinations; attempts to make an itinerary for each trip; has difficulty determining the mileage for different sections of both trips.

0 The student does not show any understanding of how the map scale can be used to plan either itinerary or how to determine distance on the map.

5 | MAPS

Wrap It Up

Vacation Time

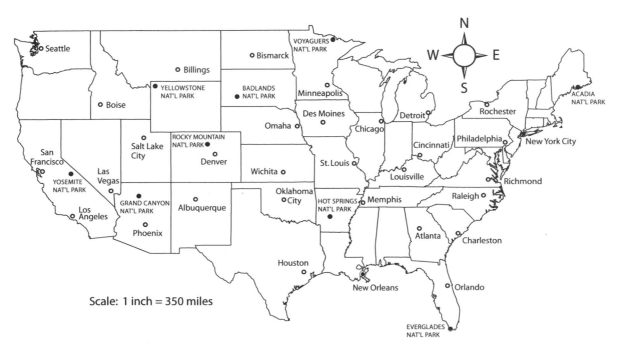

Scale: 1 inch = 350 miles

Part 1

Suppose you won an all-expense paid trip to visit a national park. Mark the national park that you would most like to visit in the continental United States. You will spend a week at the park. Then you will tour the United States for a week telling people about the park (all expenses paid, too).

Plan a round-trip tour between 500 mi and 1,000 mi, using the national park as your starting and ending place. Draw your route on the map to scale. Write your itinerary, including the number of miles between points of interest. Show how you figured the mileage.

Part 2

Your trip was such a success that the park director wants you to plan another week-long tour to talk about the park. This time, you must visit at least 5 different cities or towns. Plan a round-trip tour that is no more than 1,500 mi, using the national park as your starting and ending place. Write an itinerary for this trip. Include the number of miles between cities or points of interest. Show how you figured the mileage.

6 | BLUEPRINTS

Introduction

The term *blueprint* originally referred to a drawing that had been transferred to light-sensitive blue paper to create a white diagram on a blue background. Today, the term is often used to mean any set of detailed directions, visual or verbal. In this section, a blueprint is a scale drawing used to guide the building of an object.

Understanding Blueprint Problems

A blueprint, like any enlargement or reduction, is a scale drawing. However, unlike scale drawings, which give a scale, blueprints give actual measures. Since they are practical devices, used to help in construction, they generally provide the user with all the actual measurements needed.

Students should be aware of how blueprints are similar to and different from other scale drawings.

- Both are drawn to a scale; that is, the drawing is proportional to the actual object.

- Both can be used to find a missing dimension.

- Both can be used to make an enlargement or a reduction.

- Scale drawings provide a scale; blueprints provide actual measures.

- A scale can be determined for a blueprint, using the actual measures.

- Actual measures can be determined from a scale drawing, using the scale.

As they prepare to solve blueprint problems, students should ask themselves:

- What question am I trying to answer?

- How is the drawing related to the size of the actual object it represents?

- How does the blueprint help me picture the object?

- If I adjust one part of the blueprint, how will that affect the rest of it?

- How does the blueprint help me decide what materials and how much of each material I need in order to build the object?

Solving Blueprint Problems

Students need to realize that since blueprints are intended to guide construction, more than one drawing may be required. One or more front elevations (a view that is at a right angle to the side of the object designated as the front) usually gives only partial information about the shape of an object. Top elevations can then be included to make it easier to visualize the object. Blueprints of buildings usually have at least one top elevation per floor. Even simple objects may require more than one view.

Example:

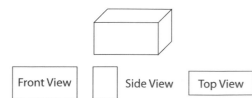

Construction of the figure shown above would not be possible with only a top view or a side view. However, since all faces of the figure are rectangles, two views are sufficient (front and side or front and top). Students may need some practice in using two such views to visualize the three-dimensional object.

In general, students will be required to analyze blueprints for the amounts of materials needed for construction.

Example: This blueprint provides all measures needed to build a skateboard ramp. The ramp is 15 ft wide. If the ramp is made from concrete but the back and base are made from wood, about how many feet of wood are needed?

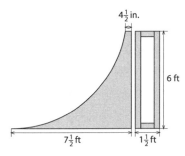

The back is a rectangular prism, $1\frac{1}{2}$ ft $\times$ 6 ft $\times$ 15 ft. Its surface area is

$2 \times [(1\frac{1}{2} \times 6) + (6 \times 15) + (1\frac{1}{2} \times 15)] = 243$ ft^2.

The base of the ramp is $7\frac{1}{2}$ ft $\times$ 15 ft. Its area is 112.5 ft^2.

The total amount of wood needed is
$243 + 112.5 = 355.5$ ft^2.

Assessment

✓ **Informal Assessment** A suggestion for informal assessment will be found on each Try It Out, Stretch Your Thinking, and Challenge Your Mind page. The recommended question will help focus students' attention on one part of the problem-solving process.

Assessment Rubric An assessment rubric is provided for each Wrap It Up. Students' completed work may be added to their math portfolios.

Thinking About Blueprints

These problems help students use their number sense and estimation skills as preparation for solving blueprint problems. Present one problem a day as a warm-up. You may choose to read the daily problem aloud, write it on the board, or create a transparency.

1. **The blueprint for a box shows only one view, which is a square. What can you tell about the shape?**
   (It's a cube since all views are the same.)

2. **The surface area of a box is 150 in.^{2} Alicia is building a box with twice the height, length, and width. Explain why the surface area will be more than double the original.**
   (You can use the *make it simpler* strategy. The surface area of a 1 in. cube is 6 in.^{2} The surface area of a 2 in. cube is 24 in.^{2}, 4 times as great. The surface area of the cube will be 4×150 in.^{2})

3. **A store's rectangular ceiling is 28 ft long and 18 ft wide. Ceiling panels are 3 ft by 2 ft. Mr. Yu wants to cover the ceiling without cutting any panels. How many panels are needed?**
   (84; $28 \div 2 = 14$, $18 \div 3 = 6$, $6 \times 14 = 84$)

4. **It took 1 qt of paint to paint a scale model of a home. If the ratio of the dimensions of the model to the actual house is 1 inch:1 ft, will 12 gal be enough paint for the real house?**
   (No; 1 inch:1 ft = 1:12, so for the area this is 12 x 12, or 144 qt or 36 gal.)

5. **A blueprint shows a circle cut in a rectangular board. The diameter of the circle is 6 in. The distance between the top and bottom of the circle and the top and bottom sides of the board is 4 in. The distance between the circle and the left and right sides of the board is also 4 in. Is a square piece of wood 12 in. on a side large enough for the board?**
   (No; $2 \times 4 + 6 = 14$ in. on a side. Students may find it helpful to sketch the blueprint.)

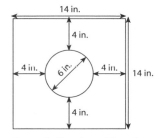

Sample Student Sketch

6 | BLUEPRINTS

Solving Problems with Blueprints

In this section, students will use their understanding of ratio, proportion, and scale drawings to solve a problem involving a blueprint. Estimating or making a table may help in finding the exact solution.

Using the Four-Step Method

Find Out
- ○ The problem asks students to locate the center of the new entrance.
- ● The distance can be calculated from the information on the diagram. ($10 - 1\frac{7}{8} = 8\frac{1}{8}$ in.)
- ○ $1\frac{7}{8} : 8\frac{1}{8}$; this means that the center is $8\frac{1}{8} \div 1\frac{7}{8}$, or almost 5 times farther from the bottom than from the top.
- ● A ratio of 4 means that the first term is 4 times the value of the second term, 1 (which is not written).
- ○ Students use the ratio 4 and the *guess and check* strategy to estimate the location on the blueprint of a point so that the distance below it is greater than the distance above it. They can then ask themselves whether the distance below appears to be about 4 times the distance above.

Make a Plan
- ● Students can make use of a combination of strategies, including *guess and check* and *make a table*.

Solve It
- ○ A possible solution path involves making a table.

	1st Guess	2nd Guess	3rd Guess
Lower Distance	5 inches	3 inches	2 inches
Upper Distance	5 inches	7 inches	8 inches
Ratio	1:1	7:3	8:2 or 4:1

As they work on the solution, ask students to explain how they are choosing values. Students should be able to explain how they are determining whether a guess is correct and how they are using each guess to improve the subsequent one.

Look Back
- ● Ask students to locate the actual position of the entrance that Mark will make. Have them compare it to the location of their estimate. Discuss whether students think estimating helped them solve the problem.
- ○ Students can check their solutions by using a proportion that includes the ratio of 4:1 and the measurements of the point they have chosen. If the ratios are consistent, they should be equal.

6 | BLUEPRINTS

Solving Problems with Blueprints

Mark is building experimental birdhouses to hang in his backyard. The blueprint that he is using is for a bluebird nesting house. It includes the front view shown here.

For his first experiment, Mark wants to lower the entrance so that the ratio of the distance below the center of the entrance to the distance above the center of the entrance will be 4. Where should Mark place the entrance?

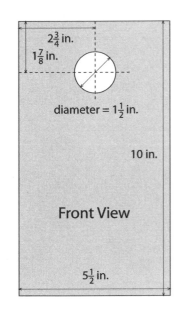

$2\frac{3}{4}$ in.

$1\frac{7}{8}$ in.

diameter = $1\frac{1}{2}$ in.

10 in.

Front View

$5\frac{1}{2}$ in.

Find Out	○ What question must you answer to solve the problem?
	• Why doesn't the blueprint give the distance from the center of the entrance to the bottom?
	○ In the blueprint, what is the ratio of the distance above the entrance to the distance below the entrance? What does this ratio mean?
	• What does a ratio of 4 mean?
	○ How can you estimate where Mark should place the entrance?
Make a Plan	• Once you understand the meaning of the ratio 4, how will you use its meaning to locate Mark's new entrance to the birdhouse?
Solve It	○ Complete the solution using the plan that you have made. Keep a record of your work.
Look Back	• How does your solution compare to your estimate?
	○ How would you check your solution?

6 | BLUEPRINTS
Try It Out

1. Find Out To help students understand the problem, ask:

- What does the problem ask you to find? (the greatest area of a square playhouse that can fit between the two sets of posts)

- What dimensions of the blueprint do you need to use? (the horizontal ground distance between the two posts)

- What is the relationship between the number of square feet and the number of square inches in a given area? (number of square inches = 144 × number of square feet)

Make a Plan To help students decide on a strategy to follow, ask:

- How will you use the blueprint to determine the area of the largest possible playhouse? (Find the distance between the two posts and use it with the formula for the area of a square.)

Solution Path

57 ft^2 of carpeting would be needed to cover the playhouse floor.

Find the distance between the two sets of posts: $112\frac{1}{2} - 22\frac{1}{2} = 90$ in.

Find the area of a square with a side of 90 inches: $90 \times 90 = 8{,}100$ in.^{2}; $8{,}100 \div 144 = 56.25$ ft^2

---

2. 11 ft of sheet metal; $127.5 \div 12 = 10.625$ ft

3. Carol has underestimated by $6.00; there are 8 rungs (no rung at the top), so total length needed = $(2 \times 8) = 16$ ft; $16 \times \$2.25 = \$36.00$; $\$36.00 - \$30.00 = \$6.00$.

Students can draw a diagram to help solve the problem.

4. 36 inches; both posts need to be increased by same amount, 84 in. − 72 in. = 12 in., 24 in. + 12 in. = 36 in.

5. $45.00; there are 10 rungs so total length needed = $(2 \times 10) = 20$ ft; $20 \times \$2.25 = \$45.00$

✓ **Informal Assessment** Ask: How far is it now from the end of the slide to the ground? (24 in.)

6. Students will need to choose a scale in order to accurately show what the rungs will look like. Ask students to explain how they decided where to place the rungs in their diagrams.

7. a. Carol could have drawn horizontal lines, one from the top of the low post to the tall post, one from the end of the slide to the low post. She would have a pair of similar triangles.

b. First she can check that each length given for the slide is correct. Then she can form proportions using corresponding side lengths and cross multiply to verify that the products are equal.

$$\frac{12}{60} = \frac{22.5}{112.5} = \frac{25.5}{127.5}$$

6 | BLUEPRINTS

Try It Out

1. Carol built the slide shown at the right in her backyard. Now she wants to build a square playhouse under the slide between the two sets of posts. If she builds the largest possible playhouse, how many square feet of carpeting will she need to cover the playhouse floor?

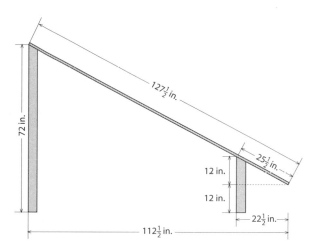

---

2. Carol made the actual slide from a piece of sheet metal. The metal is sold in pieces that are a whole number of feet long. What is the least number of feet she needed to buy?

3. The wood for the ladder rungs costs $2.25 per ft. The ladder rungs are 2 ft wide and 8 in. apart. The first rung is 8 in. above the ground. Carol estimated that she would spend $30 on wood. By how much, if any, did she underestimate or overestimate the cost?

4. Carol is considering raising the height of the slide to 84 in. If she keeps the slide the same length and at the same angle, what height will she need to make the short post?

5. What if Carol raises the height of the slide to 84 in. How much would the rungs for the ladder cost if the top rung is 4 in. below the top of the slide?

6. On a separate sheet of paper, trace the outline of the slide. Choose a scale that is appropriate to the blueprint. Then, use the information given in problem 5 to draw the ladder.

7. Carol created a pair of similar triangles on the blueprint to check that the length of the slide is correct.

   a. Describe what Carol could have done to create similar triangles.

   b. How can Carol use the triangles to check the length of the slide in the blueprint?

6 | BLUEPRINTS

Stretch Your Thinking

1. Find Out To help students understand the problem, ask:

- What question does the problem ask you to answer? (What will be the length and width of the porch after doubling its area?)

- What restriction is placed on your solution? (The porch cannot extend outside the great room.)

- What relationship do the length and width of the porch have to the rest of the house that will help you find the original measures? (The width of the porch is the same as the width of the great room; the depth of the porch can be determined by combining the depths of the great room, deck, and garage and subtracting that total from the total depth of the house.)

Solution Path

The dimensions of the new porch will be
15 ft 2 in. × 23 ft

The width of the current porch is equal to the width of the great room, 15 ft 2 in.

The length of the current porch is
76 ft − (9 ft 8 in. + 33 ft 4 in. + 21 ft 6 in.) = 11 ft 6 in.

To double the area of the porch without changing its width, double its length:
11 ft 6 in. × 2 = 23 ft

To check, compare the area of the old and new porches.
Old porch:
15 ft 2 in. × 11 ft 6 in. = 182 in. × 138 in. = 25,116 in.^{2}
New porch:
15 ft 2 in. × 23 ft = 182 in. × 276 in. = 50,232 in.^{2}
Compare: 25,116 in.^{2} × 2 = 50,232 in.^{2}

2. at least 12,920 ft^2;
width = 37 + 30 + 30 = 97 ft,
length = 76 + 30 + 30 = 136 ft,
area = 97 × 136 = 13,192 ft^2

3. 2 gal; 8 × 54 ft = 432 ft^2 of wall space to cover,
432 ft^2 ÷ 400 ft^2 = 1.16, so 2 gal will be needed

4. approximately 769 ft^2;
(15 ft 2 in. × 9 ft 8 in.) + (15 ft 2 in. × 11 ft 6 in.) +
(20 ft 10 in. × 21 ft 6 in.) = (182 in. × 116 in.) +
(182 in. × 138 in.) + (250 in. × 258 in.) = 21,112 in.^{2} +
25,116 in.^{2} + 64,500 in.^{2} = 110,728 in.^{2} ≈ 769 ft^2

5. $4,350; using the calculations from problem 1,
the area added will be 25,116 in.^{2} or 174 ft^2;
174 × $25 = $4,350

☑ **Informal Assessment** The Jantzens want to add a porch along the length of the great room. If the porch will be 10 ft wide, how much will it cost to build? ($8,333.25; the porch will be $333\frac{1}{3}$ ft^2 × $25)

6 BLUEPRINTS

Stretch Your Thinking

Deck
15 ft 2 in.
×
9 ft 8 in.

Great Room
15 ft 2 in. × 33 ft 4 in.

Covered Porch

Foyer

37 ft

Library
10 ft 4 in.
×
10 ft 4 in.

Bath

Kitchen
9 ft 2 in. × 17 ft 10 in.

DW

REF

Garage
20 ft 10 in. × 21 ft 6 in.

Bedroom
10 ft 4 in.
×
10 ft 4 in.

Bedroom
11 ft 4 in.
×
10 ft 4 in.

Bath

Master Bedroom
11 ft 4 in. × 16 ft

76 ft

1. The Jantzen family is building the house shown in the floor plan, but they want to make a few changes. They want to double the size of the covered porch without having it be wider than the great room. After its area is doubled, what will the dimensions of the porch be? You can ignore the thickness of the walls.

2. The Jantzens want to have at least 30 ft of land on each side of the house. Including the garage as part of the house, what is the least number of square feet of land they will need on their lot?

3. Mrs. Jantzen wants to estimate the number of gallons of paint that they will need for the master bedroom walls. The ceilings are 8 ft high and 1 gal of paint covers about 400 ft². How many gallons of paint should she order?

4. The builder told the Jantzens that their house contains 1,648 ft² of living space. Approximately how many square feet of space included in the blueprint are not considered living space?

5. Construction of the porch costs $25 per ft². How much does doubling the area of the porch add to the building costs?

6 | BLUEPRINTS

Challenge Your Mind

1. **Find Out** To help students understand the problem, ask:

- What question must you answer to solve the problem? (What is the total volume of the house in cubic feet, including both the living space and attic?)

- What polygons are included in the sides of the house? (rectangles and triangles, or pentagons if the front and back of the house are thought of as one polygon)

- What space figures (solids) are included in the house? (The living area is a rectangular prism; the attic is a triangular prism.)

Make a Plan To help students decide on a strategy to follow, ask:

- What plan can you use to find the volume of the house? (Possible answer: Find the volume of the living area and attic separately. Then combine the volumes to find the total volume of the house.)

You may wish to review the formula for volume of a prism with the class: V = area of base $\times$ height.

Solution Path

The volume of the house is 12,000 ft^3.

Find the volume of the living space in the house:
Volume of house space =
(10 ft $\times$ 24 ft) $\times$ 40 ft = 9,600 ft^3

Volume of attic space =
(24 ft $\times$ 5 ft) $\div$ 2 ft = 60 ft^2 $\times$ 40 ft = 2,400 ft^3

9,600 ft^3 + 2,400 ft^3 = 12,000 ft^3

☑ **Informal Assessment** Ask: What is the relationship between the volumes of the living space and the attic? How could you have used this relationship to find the total volume of the house? (The volume of the attic is $\frac{1}{4}$ the volume of the living space.)

2. $154; (2 $\times$ 10 ft $\times$ 40 ft) + (2 $\times$ 10 ft $\times$ 24 ft) or
1,280 ft^2 need painting,
2 coats $\times$ 1,280 ft^2 = 2,560 ft^2,
2,560 ft^2 $\div$ 400 ft^2 $\approx$ 6.4 gal needed,
so 7 gal $\times$ $22 = $154

6 | BLUEPRINTS

Challenge Your Mind

1. An architectural firm has drawn this set of plans for a simple cottage. All dimensions needed to construct the frame of the house are included in the drawings.

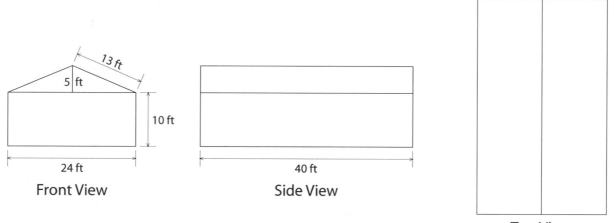

Front View   Side View   Top View

   In order to plan for their heating and air conditioning needs, the family that is considering the house wants to know the volume of space contained within the house. They know they must take into account both the living space and the attic space to find the total number of cubic feet of space. How many cubic feet of space are there in all?

2. The outside of the house, but not the roof, will need to be painted. One gallon of paint covers approximately 400 ft² of surface area. If two coats of paint will be needed, and paint costs $22 per gal, estimate the cost of the paint needed for the house.

6 | BLUEPRINTS

Wrap It Up

Design a Hangar

Ask students to compare the various dimensions of the passenger plane with distances or lengths that they know. Students may be surprised to find that the plane is probably taller than their school and longer than their auditorium or cafeteria.

- Can a blueprint or set of plans be drawn any size you choose? (Yes; as long as the ratios of dimensions are appropriate, plans can be drawn as large or small as needed.)

- Describe some of the ratios in the set of plans for the airplane. (The plane is approximately 3 times as long as the width of the tail wings; the width of the plane is approximately 3 times its height, the width of the tail wings is approximately twice the distance between the landing gear.)

Discuss what the building will look like (a rectangular prism) and how students can use the given information to determine its size.

Solutions

a. No. Since blueprints are not drawn to a particular scale, there is no need for students to choose that scale.

b. Students may wish to sketch the two planes side by side before beginning to draw the hangar. The front view of the hangar should show a width of approximately 452 ft (2 × 171 ft + 50 ft + 2 × 30 ft). The side view of the hangar should show a depth of approximately 269 ft (209 ft + 2 × 30 ft). Both views should show a height of approximately 76 ft (56 ft + 20 ft).

c. Since the building is a rectangular prism, four elevations plus a top view usually would be drawn. However, students can assume that the opposite sides of the building are the same and that the top view is a rectangle. Therefore, two views are sufficient for construction. You may wish to have students model the figure using a shoe box. From any two of the three views, it is possible to construct the box.

Assessment Rubric

3 The student demonstrates an understanding of blueprints by explaining why a particular scale is not necessary for accurately drawing them, correctly uses the given information to draw an accurate set of blueprints for the hangar using either two or three views, and explains why two views would be sufficient.

2 The student demonstrates an understanding of the concept that blueprints do not require a scale, draws a reasonable approximation of at least two views of the hangar using the given information, and with assistance can explain why two views are sufficient.

1 The student has difficulty explaining why the scale of the blueprints for the hangar need not be related to those of the plane, with help can draw a reasonable approximation of the hangar but cannot accurately apply the given restrictions and does not understand why two views are sufficient.

0 The student cannot explain the relationship between the blueprints for the plane and the size of any blueprints for the hangar, cannot apply the given information to drawing blueprints, and does not understand why two views are sufficient for a rectangular prism.

6 | BLUEPRINTS

Wrap It Up

Design a Hangar

Here are three views of a passenger plane.

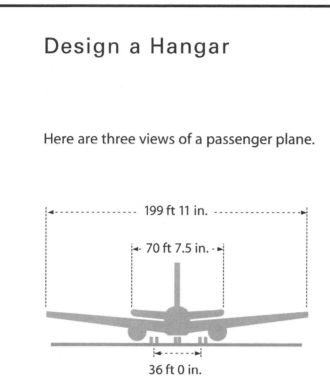

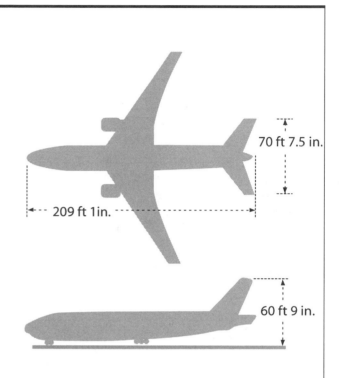

Suppose you want to make a blueprint for a rectangular hangar, a building for airplanes. The hangar must be large enough to hold two planes side by side facing the same direction. The following conditions must be met:

- There must be at least 50 ft between the planes.

- There must be at least 30 ft between a plane and a wall.

- There must be at least 20 ft between the tallest part of a plane and the ceiling.

- Both planes must be able to enter or exit the hangar at any time.

a. Do you need to use the same scale for the hangar that was used for the planes? Explain your thinking.

b. Use a separate sheet of paper. Draw views for a blueprint of the smallest hangar that will meet the above conditions.

c. How many views did you need? Why was that enough?

Class _____

Student Progress Chart

Students' Names

Section 1														
Thinking About…														
Solving Problems														
Try It Out														
Stretch Your Thinking														
Challenge Your Mind														
Wrap It Up														
Section 2														
Thinking About…														
Solving Problems														
Try It Out														
Stretch Your Thinking														
Challenge Your Mind														
Wrap It Up														
Section 3														
Thinking About…														
Solving Problems														
Try It Out														
Stretch Your Thinking														
Challenge Your Mind														
Wrap It Up														
Section 4														
Thinking About…														
Solving Problems														
Try It Out														
Stretch Your Thinking														
Challenge Your Mind														
Wrap It Up														
Section 5														
Thinking About…														
Solving Problems														
Try It Out														
Stretch Your Thinking														
Challenge Your Mind														
Wrap It Up														
Section 6														
Thinking About…														
Solving Problems														
Try It Out														
Stretch Your Thinking														
Challenge Your Mind														
Wrap It Up														